From 'Carbon Democracy' to 'Climate Democracy'?

What are the democratic requirements for effective climate action? How can 'climate democracy' be conceptualised?

Liberal democracies emerged on the back of fossil fuels, creating what Tim Mitchell called 'carbon democracy'. Three decades of climate policy have affirmed the controlling influence of fossil fuel interests. Runaway climate change now threatens the very foundations of social life. Today we face a very clear democratic question, of whether the fossil fuel sector has the right to determine the planet's climate future. Achieving global energy transformation at the scope and scale needed requires a democratic transformation, to overcome the stranglehold. This book examines these requirements. It debates the political constituencies, agendas and institutions that are emerging from climate crisis, comparing evidence of emergent themes. New claims are emerging, for 'green deals', 'climate justice', 'energy justice', 'energy democracy' and 'de-growth', reflecting a new intensity of contestation as climate change impacts deepen.

This book will be of great relevance to students, researchers and policymakers with an interest in comparative politics, democracy studies, climate change and environmental policies. The chapters in this book were originally published as a special issue of *Globalizations*.

James Goodman is Professor in Social and Political Sciences at the Faculty of Arts and Social Sciences at the University of Technology, Sydney where he researches climate change and energy transitions. He is the co-author of the book *Beyond the Coal Rush and Decarbonising Electricity* with Tom Morton.

Tom Morton is at Martin Luther University Halle-Wittenberg in Germany. He researches climate communication and is an award-winning documentary producer. He has co-authored the book, *Beyond the Coal Rush and Decarbonising Electricity* with James Goodman.

Rethinking Globalizations

Edited by **Barry K. Gills**, *University of Helsinki, Finland* and **Kevin Gray**, *University of Sussex, UK.*

This series is designed to break new ground in the literature on globalization and its academic and popular understanding. Rather than perpetuating or simply reacting to the economic understanding of globalization, this series seeks to capture the term and broaden its meaning to encompass a wide range of issues and disciplines and convey a sense of alternative possibilities for the future.

Challenging Authoritarian Capitalism
The Transformative Power of the World Social Forum
Edited by Thomas Wallgren, Uddhab Pd. Pyakurel, Catalina Revollo Pardo and Teivo Teivainen

The Reconfiguration of Twenty-first Century Latin American Regionalism
Actors, Processes, Contradictions and Prospects
Edited by Rowan Lubbock and Ernesto Vivares

Capitalism, Coronavirus and War
A Geopolitical Economy
Radhika Desai

The Rise of Green Extractivism
Extractivism, Rural Livelihoods and Accumulation in a Climate-Smart World
Natacha Bruna

Globalization, Urbanization, and Civil Society
A Non-Western Critique
Bagoes Wiryomartono

Capital Redefined
A Commonist Value Theory for Liberating Life
S. A. Hamed Hosseini and Barry K. Gills

From 'Carbon Democracy' to 'Climate Democracy'?
James Goodman and Tom Morton

Pink Tides, Right Turns in Latin America
Charmain Levy and Manuel Larrabure

The Far-Right in World Politics
Alexander Anievas and Richard Saull

For more information about this series, please visit:
https://www.routledge.com/Rethinking-Globalizations/book-series/RG

From 'Carbon Democracy' to 'Climate Democracy'?

Edited by
James Goodman and Tom Morton

Routledge
Taylor & Francis Group
LONDON AND NEW YORK

First published 2025
by Routledge
4 Park Square, Milton Park, Abingdon, Oxon OX14 4RN

and by Routledge
605 Third Avenue, New York, NY 10158

Routledge is an imprint of the Taylor & Francis Group, an informa business

British Library Cataloguing in Publication Data
A catalogue record for this book is available from the British Library

ISBN13: 978-1-032-76494-8 (hbk)
ISBN13: 978-1-032-76495-5 (pbk)
ISBN13: 978-1-003-47873-7 (ebk)

DOI: 10.4324/9781003478737

Typeset in Minion Pro
by Newgen Publishing UK

Publisher's Note
The publisher accepts responsibility for any inconsistencies that may have arisen during the conversion of this book from journal articles to book chapters, namely the inclusion of journal terminology.

Contents

PART III
Mobilisation

Citation Information

The chapters in this book were originally published in the *Globalizations*, volume 20, issue 8 (2023). When citing this material, please use the original page numbering for each article, as follows:

Introduction

'Rage into action': from carbon democracy to climate democracy?
James Goodman and Tom Morton
Globalizations, volume 20, issue 8 (2023), pp. 1257–1276

Chapter 1

Reluctant transformers or reconsidering opposition to climate change mitigation? German think tanks between environmentalism and neoliberalism
Dieter Plehwe
Globalizations, volume 20, issue 8 (2023), pp. 1277–1295

Chapter 2

Energy transition and dialectics: tracing discursive resistance to coal through discourse coalition in India
Rajshri Shukla and Pradip Swarnakar
Globalizations, volume 20, issue 8 (2023), pp. 1296–1311

Chapter 3

Democratizing global climate governance? The case of indigenous representation in the Intergovernmental Panel on Climate Change (IPCC)
Pedram Rashidi and Kristen Lyons
Globalizations, volume 20, issue 8 (2023), pp. 1312–1327

Chapter 4

Who wins and who loses from renewable energy transition? Large-scale solar, land, and livelihood in Karnataka, India
Devleena Ghosh, Gareth Bryant and Priya Pillai
Globalizations, volume 20, issue 8 (2023), pp. 1328–1343

Chapter 5

Legitimizing energy transitions through community participation: Germany and Australia at a crossroad
Franziska Mey, Anne Kallies, John Wiseman and Mitchel Watson
Globalizations, volume 20, issue 8 (2023), pp. 1344–1362

Chapter 6

Comparing local energy conflicts in NSW Australia: moving to climate generosity
Jonathan Paul Marshall
Globalizations, volume 20, issue 8 (2023), pp. 1363–1379

Chapter 7

Climate camps and environmental movements. Impacting the coal industry and practicing 'system change'
Katja Müller
Globalizations, volume 20, issue 8 (2023), pp. 1380–1392

Chapter 8

Climate movements in Germany, India, and Australia: dynamics of transition, transformation, and emergency
James Goodman and Tom Morton
Globalizations, volume 20, issue 8 (2023), pp. 1393–1410

Chapter 9

Climate, violence, resource extraction and ecological debt: global implications of an assassination on South Africa's coal mining belt
Patrick Bond
Globalizations, volume 20, issue 8 (2023), pp. 1411–1425

Notes on Contributors

Patrick Bond, professor of sociology at the University of Johannesburg since mid-2021, was born in Ireland in 1961, studied economics at Swarthmore College, finance at the Wharton School and geography at Johns Hopkins University in Baltimore. He also worked at the Institute for Policy Studies in Washington, DC, learned politics in the U.S. and took part in anti-apartheid, labour, student and community movements. After PhD research in Zimbabwe, during the early 1990s he lived mainly in Johannesburg and worked as a researcher within urban social movements. In 1994 and 1996, Patrick was a policy drafter in the Reconstruction and Development Ministry in President Nelson Mandela's office, responsible for the democratic government's first White Paper and other policies. His best-known work is *Elite Transition: From Apartheid to Neoliberalism in South Africa*, and he also authored or edited four books on climate change – including *Politics of Climate Justice: Paralysis above, Movement below* – and hosted South Africa's civil society meetings during the 2011 UN Framework Convention on Climate Change summit in Durban. He has worked at the Universities of KwaZulu-Natal, Witwatersrand and Western Cape.

Gareth Bryant is Senior Lecturer and Australian Research Council DECRA Fellow in the Department of Political Economy at the University of Sydney.

Devleena Ghosh is Honorary Professor in the School of Communications at the University of Technology, Sydney.

James Goodman is Professor in Social and Political Sciences at the Faculty of Arts and Social Sciences at the University of Technology, Sydney where he researches climate change and energy transitions. He is the co-author of the book *Beyond the Coal Rush and Decarbonising Electricity* with Tom Morton.

Anne Kallies is a senior lecturer in law at RMIT University Australia. Anne researches in the area of energy transition and climate law, with a special focus on renewable energy and electricity markets, network regulation and legal frameworks for infrastructure resilience and microgrids. Her research draws on her study and professional experience in Australia and Germany. She holds a German law degree, an LLM and a PhD, both completed at Melbourne Law School.

Kristen Lyons is a public intellectual with over 20 years of experience in research, teaching and service that delivers national and international impacts on issues that sit at the intersection of sustainability and development, as well as the future of higher education. Trained as a sociologist, Kristen works in transdisciplinary teams to deliver socially just outcomes, including for Indigenous communities. Kristen works regularly in Uganda, Solomon Islands and Australia, and her work is grounded in a rights-based approach – i.e. centring the rights and interests of local communities/Indigenous peoples, in her approach to research design, collaboration, and impacts and outcomes.

Franziska Mey is a research associate in the Energy Transitions and Public Policy Group at the Institute for Advanced Sustainability Studies (IASS) in Potsdam. Her research focuses on socio-political issues in the context of the energy transition. In addition to her work at the IASS, Franziska Mey is an honorary board member of the non-governmental organization Community Power Agency. Prior to joining the IASS, Franziska Mey worked at the Institute for Sustainable Futures in Sydney, Australia, and the World Wide Fund for Nature (WWF).

Dieter Plehwe is a senior research fellow at the Berlin Social Science Center and Privatdozent at University of Kassel Department of Political Science. He obtained his PhD (Dr Phil) in political science at Philipps-University Marburg in 1997 for the thesis titled "Deregulation and transnational integration in North America". He has worked at the WZB since 1996 in different units, including the Department Inequality and Social Policy, the Project Group Modes of Economic Governance and the Department Organization and Internationalization. He also held guest professorships and lecturer positions in Vienna and Yale University as well as research positions at the Harvard Center for European Studies and the International Center of Advanced Studies at NYU. His main research interests include private authority and neoliberalism, contested politics around climate change and socio-economic transformations. Dieter has recently published the edited volume *Nine Lives of Neoliberalism* (London: Verso, with Quinn Slobodian and Phil Mirowski) and the 'Online Dossier Lobbyismus' of the German Federal Agency for Political Education. He serves as one of three main editors of the Journal *Critical Policy Studies* and has developed an online repository and blog on think tank networks (www.thinktanknetwork.net).

Jonathan Paul Marshall is an anthropologist who studies technology, the social psychology of climate change, disruption and disorder. He is the author of *Living on Cybermind: Categories Communication and Control* and co-author of *Disorder and the Disinformation Society: The social dynamics of information, networks and software*. He edited *Depth Psychology, Disorder and Climate Change* and co-edited *Environmental Change and the World's Futures: Ecologies, Ontologies, Mythologies* and *Earth, Climate, Dreams: Dialogues with Depth Psychologists in the Age of the Anthropocene*. He has also co-edited special issues of *Globalizations*; *The Australian Journal of Anthropology*; *Energy Policy* and *Energy Research and Social Science*.

Tom Morton is at Martin Luther University Halle-Wittenberg in Germany. He researches climate communication and is an award-winning documentary producer. He has co-authored the book, *Beyond the Coal Rush and Decarbonising Electricity* with James Goodman.

Katja Müller, Dr. Habil., is a social anthropologist with research interests in energy humanities and environmental anthropology, as well as visual, digital and museum anthropology. She has been with the Leipzig ethnographic museum for several years and has been working with IASS Potsdam, Halle University, UTS Sydney and UCL London on digitization, coal mining, renewable energy, and transformation. Her latest books are *Beyond the Coal Rush* (2020) and *Digital Archives and Collections* (2021).

Priya Pillai is PhD Candidate in the School of Communications at the University of Technology, Sydney.

Pedram Rashidi worked for over 12 years in the Iranian and Australian energy industries before commencing a PhD at the University of Queensland (UQ) on global environmental governance. His thesis combined his experience in the renewable energy sector with political theory and the sociology of knowledge. At UQ he researches environmental governance, development and climate politics.

Rajshri Shukla is a PhD candidate in sociology at the Department of Humanities and Social Sciences, Indian Institute of Technology Kanpur, India. For her doctoral thesis, she is working on a comparative study between India and Australia, revolving around the issues of energy justice within the process of energy transition away from fossil fuels. Her work has appeared in *Environmental Communication*.

Pradip Swarnakar is Professor of sociology at the Department of Humanities and Social Sciences, Indian Institute of Technology Kanpur, India and Adjunct Fellow, Social and Political Sciences Programme, University of Technology Sydney, Australia. He has been a visiting scholar at the University of San Francisco, University of Helsinki and Helmholtz Centre for Environmental Research GmbH-UFZ. He is an associate editor of the social science section of *Heliyon*. He is co-editor of the book *Bottom-up Approaches in Governance and Adaptation for Sustainable Development: Case Studies from India and Bangladesh* (SAGE Publications, 2017).

Mitchel Watson is a PhD researcher in system innovation, ecological economics and human geography at RMIT University Australia and is Director of Programs and Recovery, Rural and Regional Victoria at the Department of Jobs, Precincts and Regions, Victoria State Government. Mitchel is recently retired from the board of the Hepburn Wind Co-operative – Australia's first community-owned large-scale renewable energy generator.

John Wiseman is a professorial research fellow at Melbourne Sustainable Society Institute; a senior research fellow with Melbourne Climate Futures and an adjunct professor at the Melbourne School of Population and Global Health, University of Melbourne. Other current roles include Visiting Fellow, Centre for Climate Economics and Policy, Australian National University and Research Fellow at the Centre for Policy Development. Prof. Wiseman's major research and policy contributions have been in the fields of climate change risks and climate justice; the political economy of globalization and social, economic and environmental policy frameworks. His current work focuses on the actions needed to accelerate the transition to a just and resilient post-carbon society.

Introduction – 'Rage into action': from carbon democracy to climate democracy?

James Goodman and Tom Morton

ABSTRACT

Overcoming fossil fuel dependence entails overcoming the constraints of 'carbon democracy'. Transition to decarbonized society poses a profound democratic question, of whether the fossil fuel sector has the right to determine the planet's climate future. With this, energy transitions to address climate change are transformed into much deeper democratic transitions. This is already in evidence, as new claims are generated, for 'climate justice', 'energy democracy', 'new green deals', 'de-growth', and more. The imperatives for transition, and the in-built barriers to it, pose real questions for how society and politics are understood. How far does climate transition force new models of 'climate democracy' onto the table? Theory-policy linkages and concrete case studies are needed to explore the dynamics of democracy under transition, and to understand shared barriers and alternatives. This introductory article draws out perspectives on these themes in the Special Issue, 'From Carbon Democracy to Climate Democracy'.

> I know people everywhere are anxious and angry. I am, too. Now is the time to turn rage into action. Every fraction of a degree matters. Every voice can make a difference. And every second counts. UN Secretary Secretary-General António Guterres (2022).

At 8.30 am on 13 April 2022, Deanna 'Violet' Coco, a 32-year-old climate activist and member of the group Fireproof Australia, drove a hired truck onto the Sydney Harbour Bridge and stopped it, blocking commuter traffic across the bridge in one lane for around an hour. Coco stood holding a lit flare and live-streaming from the top of the truck until she and three other protesters were arrested and removed by police. Coco pleaded guilty to seven charges including using or modifying an authorized explosive and resisting a police officer during arrest. It was her twenty-first arrest for climate protests. Coco said she was motivated to become involved in climate activism by the 2019–20 'Black Summer' bushfires in Australia in which 17 million hectares were burnt, 33 people were killed and 3094 homes were destroyed (Richards et al., 2020). According to the World-wide Fund for Nature, nearly 3 billion animals were killed or displaced by the fires (WWF, 2020).

Following the 2019–2020 fires a special Issue of *Nature Climate Change* addressed the political impacts for Australia, suggesting the fires were bringing home climate change to insulated elites (Editorial, 2020). In it, Lesley Head described the 'double reality' of engagement and avoidance in rich countries such as Australia, asking if climate impacts were introducing a new and irresistible dynamic into the political system (Head, 2020). Certainly, the advent of large-scale climate disruption can be closely linked with the wave of climate protest following the 2015 Paris Accord.

Growing impacts and failing policies have shaken democratic legitimacy. The upsurge in protest led by school children from 2018, mainly in the richest countries of the world, was met with blank failure, and deeper obfuscation. Disruptive climate action, post-Covid, faced ever-more repressive measures: in defending the indefensible, liberal democracies increasingly resorted to bare force, with a wave of anti-protest laws and actions against climate organizations from pre-emptive disbanding of organizations to 'exemplary' repression.

Deanna Coco was the first person to be charged under laws introduced by the NSW State government which specifically targeted climate protests and carry jail terms of up to two years and fines up to A$22,000 for blocking public roads, rail lines, tunnels, bridges, and industrial estates (Parkes-Hupton, 2022). In December 2022 Magistrate Allison Hawkin sentenced Coco to 15 months in jail for blocking the bridge and refused bail, describing her as 'emotional and childish' and said she had made 'an entire city suffer' (Kurmelovs, 2022). Sections of the media condemned her as a 'climate fanatic'. The conservative Premier of New South Wales stated that the judgement was 'pleasing to see' and would deliver a 'clear lesson'; the Labour opposition agreed. In response, the United Nations rapporteur on freedom of association stated 'peaceful protesters should never be criminalised or imprisoned'; a Climate NGO, the Green Institute, condemned the repression and called for the movement to 'start building a real democracy that works for people and planet' (Rose & McGowan, 2022).

Sydney is not alone. There has been similar legislation across several States in Australia following disruptive climate actions. Law-making ostensibly to deter direct action by climate protesters has become a global trend, geared to criminalizing and discouraging protests (Tigue, 2022). In 2022 UK members of Just Stop Oil and Insulate Britain were being held for up to six months in jail before being tried for blocking roads (Macalister, 2022). In Germany, conservative politicians compared members of Letzte Generation (Last Generation) who had glued their hands to a road in Munich to the Red Army Faction (the Baader-Meinhof Group; Die Zeit, 2022) and in 2023 the German police launched a coordinated series of raids on Last Generation (Connolly, 2023). In the USA, 15 states have passed laws making it a felony to 'impede' the operations of a pipeline or power plant (Kusnetz, 2021). The consequences for climate activists in the Global South can be much graver. In India, as we outline in our article for this Special Issue (Goodman & Morton, 2023), Fridays for Future activists have been targeted by the authorities; increasingly, anti-coal campaigners have been put under investigation and jailed, and a range of climate campaign organizations have been closed down (Shih et al., 2023). In South Africa, as outlined in Patrick Bond's contribution (Bond, 2023), repression has included the murder of 63-year-old Fikile Ntshangase who had successfully opposed the expansion of a coal mine in KwaZulu-Natal. Her assassination – still unsolved – is representative of a broader trend identified by the authors of the Global Environmental Justice Atlas which shows that environmental defenders face '… high rates of criminalization (20% of cases), physical violence (18%), and assassinations (13%), which significantly increase when Indigenous people are involved' (Scheidel et al., 2020).

Climate policy failure is leading to the increasing use of various forms of direct action, civil disobedience, disruption, and performative protest by the climate movement, and also to an increasingly repressive response from state authorities, coupled with rhetoric which seeks to marginalize and delegitimize protesters. This dynamic is symptomatic of a more fundamental problem: the failure of liberal democracy to deliver real and necessary climate action (Almeida, 2019). As UN Secretary-General António Guterres declared in a speech marking the release of Part 2 of the IPCC Sixth Assessment Report in February 2022, 'today's report is an atlas of human suffering and a damning indictment of failed climate leadership' (Guterres, 2022). The answer to failed political leadership, mostly elected under liberal democracies, was climate activism: 'Climate activists are sometimes depicted as dangerous radicals. But the truly dangerous radicals are the countries that are increasing the production of fossil fuels'. Guterres said he took 'inspiration from all those on the frontlines of the climate battle'; as we quote at the opening of this article, the world needed to turn 'rage into action'.

That appeal, directly from the ostensible peak of global normative authority, signals something of the crisis in political legitimacy. The existing structures of national political life, in particular the structures of the liberal democratic state, have been failing to produce climate leadership (Goodman & Morton, 2014). The answer to this, we argue, is more democracy, not less, as argued by some (Shearman & Smith, 2007). The appeal to activism, and to 'rage' against that failure, is an appeal to a different type of democracy, a 'climate democracy' perhaps. It is not clear yet what form that would take, beyond having a participatory aspect, centred on taking action. 'Action' can take many forms, from advocacy to insurgency, with varying degrees of democratic engagement and participation and there is a long history of social movement debate over strategy, action, and mobilization (for instance, Arrighi & Wallerstein, 1989). Each movement has its own context and logic, though arguably the context for climate movements is qualitatively different, raising very real dilemmas over what forms 'action' should take. Something of the urgency and scope of this debate is conveyed in Andreas Malm's *How to Blow Up a Pipeline* (2021), along with the debate it has stimulated, including a feature film dramatizing the issues (Goldhaber, 2023). From our perspective, the question of capture by fossil fuel interests directly demonstrates the redundancy of current democratic institutions. As Timothy Mitchell argued in the final pages of *Carbon Democracy*, 'the democratic machineries that emerged to govern the age of carbon energy seem … unable to address the processes that may end it' (Mitchell, 2013, p. 399).

Beyond carbon democracy

The contributors to this special issue all seek, in differing contexts and from differing perspectives, to address the questions that follow from Mitchell's provocation: what new 'democratic machineries' might emerge to govern the age of decarbonization, which actors will shape them, and what new forms of collective action and agency might this process require and enable? These questions pose urgent and profound challenges for social and political theory (see also Urry, 2014). The transition to a decarbonized society is not simply a technocratic issue, but a much more profound democratic question, of whether the fossil fuel sector has the right to determine the future of the planet's climate, and with it, the future of humanity. Progress on energy decarbonization remains piecemeal and tantalizingly slow: as the International Energy Agency stated in its *World Energy Outlook 2021*, 'every data point showing the speed of change in energy can be countered by another showing the stubbornness of the status quo' (IEA, 2021, p. 1). With this, the ongoing political influence of the fossil fuel sector is widely identified as the central barrier to effective policy (Influence Map, 2022). The institutional power of the sector is being laid bare. In many countries energy policy based on fossil fuels has become highly politicized, and a growing range of policy sites and alternatives have emerged.

At the same time, however, national populist movements and their leaders have seized on people's fears about the costs of decarbonization and its impacts on livelihoods and 'ways of life' as fertile ground for mobilization; one recent example is the public anxiety about the German Federal Government's 'Heating Law' (Heizungsgesetz), which the right-wing populist party 'Alternative for Germany' (Alternative für Deutschland) has exploited to boost their vote, especially in

Eastern Germany. As Huber et al. argue, citizens' support for populist parties which oppose climate action is closely linked to distrust of institutions; science is seen as just another institution, and 'climate scientists as part of the self-serving elite that betrays the people' (Huber et al., 2022). The growing authoritarian response to climate protest is itself partly a response to climate populism, and signals a shift in the centre-ground. Centrist governments adopting limited climate policy commitments have consolidated an emergent bloc of dominant forces that embrace the rhetoric of ecological modernization. A package of 'green growth', corporate renewables, offsets for carbon sinks, 'offshoring' of emissions, and 'net zero' commitments are offered 'from above'. These responses take-on the role of adapting but maintaining the status quo, instituting a 'green passive revolution' that can become authoritarian in marginalizing its critics (Mookerjea, 2019). Passive revolution cannot solve the crisis: emissions continue to rise and climate instability accelerates (Carroll, 2020). The model is vulnerable to protest from the far Right, but also from the climate movement, that has become radicalized in the face of policy failure. Anti-climate protest measures are normalized as forms of virtue-signalling to the populist right, and they increasingly dovetail with emergency measures taken in the face of climate-induced fires and floods (Darian-Smith, 2022). Here, the strong-arm state represses both the political and biophysical symptoms of a deepened crisis.

On the other side of this emerging global antagonism is the remarkable upswelling of climate movement claims. These are variously advancing 'non-reformist reforms' that, as Carroll argues, 'may hold out the possibility of an anti-passive revolution' (Carroll, 2020, p. 2). Climate policy advocacy was initially centred on meeting emissions-reduction targets. Intensifying policy failure and the rise of the global justice movement in the late 1990s forced a shift. Critiques of corporate globalism, framed as 'justice globalism' became closely articulated with the climate policy advocacy agenda (Steger et al., 2013). Critiques of neoliberal climate policy emerged, along with sharper analysis of growth-based assumptions, and associated 'de-growth' critiques (see Kallis, 2020). With this highly generative ideological antagonism, 'climate justice' claims, and advocacy for new 'green deals' in fossil-fuel dependent economies such as those of India, Australia, Germany, South Africa, and the United States, have been increasingly on the agenda. Climate movements are, in the process, creating new socio-ecological relations and frameworks, in a dynamic and dialectical process (Moore, 2015). A plethora of new activist groups and organizations have emerged: initiatives such as Fridays for Future and Extinction Rebellion, which are decentralized, highly networked, and globalized, have given new impetus and visibility to climate advocacy, creating what Beck called 'cosmopolitan events', which stage the 'global risk' of climate change (Beck, 2015, p. 260). The actions are highly mediatized and able to galvanize public awareness of the climate emergency (Haunss et al., 2019).

The ideological struggle over climate policy, and the movements it generates, are of critical importance. Writing in early 2020, before the full scale of the coronavirus pandemic had become apparent, Andreas Malm argued that over the last two decades 'the climate movement in the global North has undergone several cycles of intense activity, each on a larger scale than before', forming a 'cumulative process and rising loop' (Malm, 2021, p. 13, 21). The political momentum built over the course of 2019 was interrupted by the coronavirus pandemic; but in the wake of the pandemic, and with the failure of the Glasgow Climate Summit in 2021 to deliver meaningful progress, a number of new groups came to prominence. After three decades of climate policy-making many now argue that the 'democratic machineries' of fossil capital leave them with no choice but to engage in disruptive direct action. Some have highly specific aims, such as Just Stop Oil and Insulate Britain, others have specific constituencies, such as Scientists Rebellion. Their aims and tactics may differ from those of more mainstream climate advocacy organizations, but all are locked, in one way or another, in a contest with the entrenched power and influence of the fossil fuel industrial complex (see Goodman et al., 2020).

Challenging that power and influence requires a thorough-going renewal of democratic processes (Burnell, 2012; Fioirino, 2018). The social process of energy transition forces open issues of climate democracy, of the need to establish new socio-ecological relations for climate stability. From these flow further questions about 'potential transformational impacts' (Van Veelen, 2018, p. 644). As Mitchell documents, successive energy transitions have enabled radically different social formations, with contrasting ecological and political impacts. The expansion of coal-fired power in the Nineteenth Century engendered a mass urban citizenry capable of deepening liberal democracy. It was intertwined with the emergence of an organized labour movement capable of interrupting the extraction of coal at source, and thus able to press claims for social and economic justice, forging 'carbon democracy' (Mitchell, 2013). Oil, a fuel that was increasingly extracted in colonial or post-colonial settings remote from the industrialized economies whose growth it fuelled in the Twentieth Century, could be easily transported, and thereby weakened the power of organized labour, while enabling new forms of social organization, associated with new patterns of settlement and mobility (Mitchell, 2013). Mitchell's broader argument is that what we have come to recognize as liberal democracy is deeply imbricated with the extraction and exploitation of fossil fuels and is unthinkable without them. Liberal democracy is carbon democracy and vice versa, the two exist in a co-dependent relationship. Moving beyond the age of carbon energy also requires a fundamental reimagining of what democracy itself means. As Mitchell argues: 'the possibility of more democratic futures ... depends on the political tools with which we address the passing of the era of fossil fuel' (Mitchell, 2009, p. 423).

Today's transition away from fossil fuels is similarly establishing new socio-ecological relations of energy, driven by the underlying climate contradiction (Moore, 2015). The result is profound social antagonism, but also transformation. As Szulecki (2018) argues, across the globe 'energy governance is at a crossroads, facing inevitable change, perhaps even a 'third industrial revolution', with widescale potential for change. The transition from a centralized, mostly fossil fuel-based power sector, towards a distributed renewable energy system brings new small and medium power producers, as 'prosumers'. At the same time, technological innovation and the falling price of renewable energy and infrastructure creates new potential for social justice at the centre of decarbonization. In contrast with historic energy transitions, the current transition is consciously governed. New roles are emerging for the state in driving decarbonization through public provision, offering grounds for socialized energy and 'energy democracy' (Szulecki, 2018). Alliances emerge across social fields, centred on contending models for transition, opening up new visions and alternatives (Hess, 2018).

Climate democracy, however, is not reducible to energy democracy. In the Afterword to the 2013 edition of Carbon Democracy, Mitchell stressed that decarbonization will not inevitably lead to a more democratic world; in battles over the shape of future energy systems, the question of democracy is open, and open to contestation (Mitchell, 2013, pp. 266–7). In a more recent interview in the journal City, he defined democracy not as 'the spread of an idea ... in people's heads or a culture of accountability', but as 'the way in which certain forms of movement or arrangement or structures of the ways things are accumulated or populations are governed are open to interruption and challenge' (Abourahme & Jabary-Salamanca, 2016). Here Mitchell deliberately frames democracy not as a set of institutions or abstract political principles, but as intervention, a process of 'interruption or challenge' to existing power structures. Implicitly, this challenge comes from below, and involves the emergence of new actors and alliances of actors, and new forms of agency. The structures of the carbon-based fossil-fuel economy have formed and evolved over two centuries since the beginnings of the industrial revolution in Europe; the structures of the renewable energy economy are still in a highly dynamic process of development. As with any substantial shift in the status quo there is a powerful counterforce. Resistance is not surprising: as Mark Carney, Gov. of the Bank of England noted, 'one third of global equity and fixed income assets' are carbon-exposed, set to be 'stranded' by decarbonization (Carney, 2015). The current global energy transition began in the 1990s and was driven by technological innovation, environmental movements able to exert increasing influence in the political mainstream, such as the German Greens, and economic incentives provided by governments, especially in northern Europe. Today, renewable energy is increasingly corporate and large-scale, making decarbonization hostage to continued returns on renewable capital (McCarthy, 2015). We are interested in how capitalization and the advent of neoliberal renewable energy are open to 'interruption and challenge', and how broader forms of collective organization, action, and socialization emerge in the moment of climate crisis, ensuring wider capacity and legitimacy to overcome the status quo (see Böhm & Sullivan, 2022).

Certainly, there are grounds for change, with a multiplying range of policy sites and alternatives emerging to challenge the dominance of the fossil fuel sector. Demands are highly generative, creating a powerful nexus between renewable technology and social organization. In the wake of social movement pressure these can come 'from above', though are often compromised. The inauguration of the Biden administration for instance signalled a major shift in US rhetoric on climate policy, and a commitment to implement significant elements of HR 109 (more commonly known as the 'Green New Deal'), which would 'empower American workers and businesses to lead a clean energy revolution' and 'achieve net-zero greenhouse gas emissions through a fair and just transition for all communities and workers by 2050' (US Congress, 2019, p. 5). What the IEA calls the 'stubbornness of the status quo', however, is perhaps nowhere more starkly illustrated than by the difficult passage the Biden administration's package of climate and energy measures through the legislature. In 2022 the administration finally secured the support in Congress for a pared-back series of climate initiatives. Originally set at USD550 billion, the finally-agreed package was USD369 billion, and significantly, the measures were directed at expanding renewable energy rather than phasing out fossil fuels. Further, to secure support in Congress the measures were linked to the expansion of leases for coalmining and gas drilling, including in Alaska, condemned as a 'climate suicide pact' (Milman, 2022).

The 'stubbornness of the status quo' clearly has its origins in dominant corporate fossil fuel interests, but it is also firmly rooted in the geopolitical logic of world hierarchy. The US and other dominant 'Apex' states, the EU and Japan included, effectively hedge against renewables, taking a two-track approach of relative (not absolute) reduction in fossil fuel dependence. They compete with a range of 'second-tier' countries to gain expanded access to fossil fuel reserves, and this both reflects and accentuates rivalrous inter-imperial and sub-imperial antagonisms. Energy diplomacy, and indeed energy wars, are nothing new, though arguably are figuring more prominently, notably with Russia's gas and oil diplomacy. Geopolitical antagonism of this sort legitimizes authoritarian responses, and generates growing security anxieties and military build-ups (see Albert, 2021).

The spill-over of resource rivalry and capital-capital competition into climate policy is well-known. For decades there has been a process of bidding down and sidestepping climate policy commitments, with self-reinforcing effects in the zero-sum logic of emissions reduction. Several countries have intermittently been accused of blocking more effective global agreement-making

– from the US to China, to India, and Saudi Arabia. The US under George Bush famously exited the Kyoto Agreement in 2001, citing competitiveness concerns. More recently India led a last-minute intervention at the COP26 Glasgow climate summit which replaced a commitment to 'phasing out coal' in the final climate pact with the phrase 'phasing down coal'. Pressure from Saudi Arabia led to the repeated inclusion of references to carbon capture and storage in the IPCC's Sixth Assessment Report on Mitigation of Climate Change (Farand, 2022). And even if, as some recent commentary suggests, 'the intensifying effects of climate change potentially threaten both the stability of the country and the longevity of Chinese Communist Party rule' (Prytherch et al., 2023), China, the world's largest emitter of CO2, has consistently argued that its path to decarbonization necessitates continuing growth in emissions until 2030 (rather than the 45% reduction by 2030 enshrined in the Paris Agreement). In one sense, these actions by some of the world's largest emitters and producers of fossil fuels could be seen as challenging Western dominance of the climate policy agenda; equally, they often provide a convenient pretext for Western governments, and especially the US and Australia, to pursue less ambitious climate policies themselves, and vice versa, into the vortex of the 30-year climate security complex.

Despite these pressures, there has been something of a shift underway, at least since the 2015 Paris Summit. Rivalry to avoid emissions reductions has dominated international climate policy, with climate action seen as costing the economy and debates centring on the issue of who is responsible, and hence who is to bear the presumed decarbonization costs. Recently, though, the ground has shifted as climate action is no longer framed as a cost to present-day society to avoid future disruption, but is instead offered as a benefit, needed for prosperity and growth. This is in large part due to the 'Green Growth' agenda as climate action has moved to the centre of agendas for industrial and social change, as a new socio-ecological fix for capital accumulation. The International Renewable Energy Agency, for instance, predicts the global energy transition will cost $15 trillion and produce $98 trillion, in the process creating 7 million jobs by 2050 and substantially improving wider development and welfare outcomes (IRENA, 2020). Decarbonization is presented as bringing social benefits and opening up new potential for 'just transition' and 'energy justice' (Aranof et al., 2019; Burke & Stephens, 2017; McCauley & Heffron, 2018; Rosemberg, 2010). This lends a strong purposive aspect to the energy transition, as desirable for the economy and for society, as well as necessary for the environment. As a result, it engages an increasingly broad set of interests, actors, constituencies, and protagonists. This widening appeal directly underpins climate movements, as movements for wide-scale societal change.

At one level, the new dispensation is subsumed into geopolitical rivalries. Certainly, the rush to renewables is driving new efforts to appropriate co-called 'critical minerals', securitizing sustainability efforts. Extractivist rivalry cascades 'downwards' into competition over access to resources and the peripheralized societies that host them, accentuating resource curses (see Chagnon et al., 2022). For dominant countries the new socio-ecological fix is itself posed as an issue of economic diplomacy, and ultimately as a security challenge. It has long been predicted that the centre of gravity in the world system is soon-to or has shifted from the Atlantic to the Pacific, reflecting the growing centrality of newly-industrialized and now leading 'green' economies, especially China. From this perspective, the centrality of China to the global renewable transition is cited as a measure of its growing centrality to the future of the world system (Hafner & Tagliapietra, 2020). Great power rivalry is inter-meshed with competition between countries seeking to define leading-edge technology in the energy transition. Historically, the economic interregnum, in the transition from one mode of accumulation to the next, correlates closely with the inter-state interregnum between one set of leading powers to the next (Gills & Frank, 1994, p. 4). In this cyclical model crisis disposes of the old order and announces the new. From this perspective, the incorporation of decarbonization into state-led developmentalist agendas, and the related securitization of transition, may be interpreted as a measure of its success.

Yet the climate crisis is not the same as earlier cyclical crises of capitalism. The climate crisis signals a socio-ecological breakdown where, simply put, climate instability increases as emissions increase. It is quintessentially a cumulative crisis, not a cyclical crisis (see Ollinaho, 2023). Global greenhouse gas emissions are not wiped off the ledger with a change in the mode of accumulation. With this, the global rush to energy transition exemplifies the inherent limitations of a new mode of accumulation centred on 'green' growth (Albert, 2021; Kallis, 2020). We can speculate at the implications of this world-historic moment for the cyclical logic of the world system. Rather than an ongoing cycle, the current climate crisis suggests a profound break. We can expect imperialism to become increasingly subsumed in climate, perhaps defining a logic of domination-via-emissions, of resource peripheries, future generations, and endangered species (see Goodman, 2021). This is already reflected in some of the recent more radical injunctions to addressing climate emergency for instance as the Last Generation. From this perspective, great-power rivalry and domination, indeed the whole panoply of geopolitics, is rendered marginal.

Certainly, the logic of global climate change, and the necessary responses to it, are increasingly setting the pace of change. In this intensifying crisis climate movements have a central role to play, of continuing to force the climate issue into the international agenda and to assert the need for society-wide transformation in order to address it. Reflecting this, their role is strongest in generating ideological alternatives, as agenda-setters for climatized society. Climate change is often said to be a product of the Anthropocene, in which human activity, shaped by capitalism, becomes a 'major geological force' (Crutzen & Stoermer, 2000): the corollary to climate change is the emergence of a new form of self-aware climate agency. But, as the sociologist Mike Hulme puts it, 'we have only tentative understanding of the implications of such a new role and only limited means at our disposal to exercise purposeful, as opposed to inadvertent, agency' (Hulme, 2010, p. 1). The scope and depth of this challenge is compelling. Climate change transforms the established modes of historical agency: society has to gain political purchase over structures that exist at a scale and register far beyond the current preoccupations of political life (Chakrabarty, 2021). Climate agency thereby reconfigures established categories of human justice, well-being, and democracy. People are only beginning to imagine the implications for the political community, and for the very language of politics (Wainwright & Mann, 2013). Climate change politics is forcing new alignments and creating new constituencies and brings in a wide range of actors. The 'climatization' of politics (Aykut et al., 2019) now extends far beyond the climate movement. In this context the social and political dynamics of climate movement action become vitally important in defining the scope for transformations into the future.

Themes for climate democracy

New forms of climate democracy are emerging from the climate crisis, against the logic of carbon democracy (see LeQuesne, 2020). Here we focus on three main aspects: representation, participation and mobilization. The discussion of representation centres on ideological and discursive conflicts, especially between the fossil fuel bloc and its contenders, though it could also extend to institutional questions. Participation is discussed mainly in terms of the capacity for engagement in the energy transition and related agendas for social justice, social ownership, and energy democracy. Mobilization is addressed in terms of the dynamics of the climate movement and the emergence of wider climate constituencies and climate-engaged social movements. Across these aspects, there is an emphasis on climate democratization as a dialectical process, grounded in the contest between social forces, but open-ended in terms of the transformative agendas it generates.

Representation?

At the national level, there is an intensifying contest over the control of energy policy. The climate imperative has radically disrupted the long narrative of fossil fuel developmentalism. But unseating fossil fuels from the energy mix requires a much larger project of dismantling long-ingrained structures of state capture. As we have argued, the fossil fuel sector is heavily entwined with the project of liberal democracy. Indirect limited democracy, on the Madisonian model, was invented to insulate elite interests from the masses, so today it protects the fossil fuel elites from the necessity for their own demise (Goodman & Morton, 2014). The democratic interplay of discourses, in a field of interest groups and civil society players, is supposedly a key foundation for liberal democracy. Yet fossil hegemony is maintained through the dominant public discourse, and through defining the priorities for state power. Ironically, the negation of future democratic life thrives in liberal democracy.

Plehwe's article in this Special Issue focuses on these issues in Germany, a world leader in energy decarbonization and the 'perfect site for a case study both on "path dependencies" and "institutional change" in the area of climate change mitigation' (2023, p. 2). Despite the country's remarkable track record of energy transformation, Plehwe demonstrates how the dominant discursive field still is shaped by fossil fuel interests. In the German debates on climate and energy, he shows how transformative climate action (on the scale that is required) is positioned as an ostensibly radical outlier, against incremental 'business-as-usual' energy transition approaches and demands for continued reliance on fossil fuels. The 'radical' climate model advances transformative 'energy democracy' against the corporate-led renewables 'modernisation' approach and the pro-fossil fuel status quo. Plehwe analyses the respective 'discourse coalitions', the intersections between constituencies and the capacity for influence over the policy agenda.

Think-tanks, Plehwe finds, play a key role in these manoevres, mediating movements, interests and policy. They enable policy change, creating disruptive discourse coalitions: energy transformation arises at the nexus between social democrats and radical ecology; it is later undermined by an alliance between fossil fuel interests and neoliberals. Notably, think tanks act as 'hands-off' proxies for dominant fossil fuel interests, 'dedicated to eroding confidence in the energy transformation' (2023, p. 12). New Right neoliberalism has here found its home with fossil fuel interests, advocating 'market' solutions to unravel the 'energy transformation', driving a wedge between the 'social market' and climate action. Social democrats have buckled to neoliberalism, promoting market 'efficiency' and 'competitiveness', while insisting they act for the climate. As Plehwe argues: 'neoliberal environmental economics has ... become a dominant discourse utilized to delay the exit from fossil fuels and to block the movement towards decentralized energy production' (2023, p. 15).

Capacity to manipulate the public sphere, and thereby delimit democracy, is central to the ongoing dominance of fossil fuels. Dominance in the face of climate change has to be actively pursued to secure fossil fuel interests against the growing evidence of ineffective climate policy. A field of conflict emerges between contending discourse coalitions – for and against fossil fuels – locked into a struggle to influence government policy. In this, developmentalist narratives play a central role, whether to rescue the dream of unending fossil fuel development, or to map-out a renewable future for 'green' prosperity. In the liberal democratic Enlightenment tradition, the future is presented as a choice for democratic deliberation. Ironically, liberal democracy becomes its own grave-digger, treating the bio-physical necessity for climate action as a matter of choice. Liberal democracy as the vehicle for elite interests becomes a potential victim of its own game.

Unquestionably, as Shukla and Swarnakar highlight in their study of Indian discourse coalitions over the future of coal, a profound conflict has opened-up over fossil fuels and their place in India's developmental narrative. That conflict, they outline, has a dialectical logic driven by the escalating disruptions of climate change. Systemic climate dislocation has already arrived, and especially in India: the disjuncture between elite-defined fossil fuel-based development and the demand for climate action creates a sharpening conflict, with transformative potential. As they state: 'it becomes urgent to more systematically explore for possibilities of various avenues from where the resistance to coal expansionism can emerge' (2023, p. 3). Shukla and Swarnakar explore this 'discursive dialectical process' as the incumbent coal-centred discourse coalition is increasingly contested and confronted by an emergent renewables coalition, demanding accelerated energy decarbonization. With this, they use Bloch and Gunderson to establish a model for understanding the present-day emergence of 'merely possible solutions within the dialectics of climate change' (Shukla & Swarnakar, 2023, p. 2).

Where Plehwe explores the 'back end' nexus of fossil fuel thinktanks and their opponents, Shukla and Swarnakar address its 'front end', in terms of what voices are represented in the media and how this reflects the underlying dialectic of climate and energy. Shukla and Swarnakar's analysis focuses on statements on coal and renewable transitions reported in India's three highest-circulation newspapers. They found 229 relevant articles with more than 500 relevant statements from Paris 2015 to the Pandemic 2020. Statements are coded for source and ideological affinity across more than 40 concept categories, yielding a mapping of affiliation networks. Two main affinity blocs are highlighted – one supporting renewable transition storylines, the other backing coal. Renewable storylines appear more often in the media but are more dispersed; coal storylines are more distilled and account for two of the three most prevalent narratives: the most prevalent statement is 'Expansion of renewable energy needed'; the next most prevalent are: 'India needs coal to meet its developmental challenges' and 'Coal will stay financially viable in future' (2023, p. 8).

Shukla and Swarnakar point to the paradox that post-colonial developmentalism allows the Indian state to simultaneously acknowledge climate crisis and prioritize developmental needs, defined both in terms of expanded renewable energy and expanded coal consumption: the issue of climate change is a relegated, displaced elsewhere. The political struggle then centres on the developmental benefits and disbenefits of coal as against renewables. The question of whether renewable energy is defined as an alternative to coal, or as a supplement to it, is critical. This struggle, as Shukla and Swarnakar demonstrate, is framed by the contending developmentalist 'storylines', and their resonance and coherence. The media analysis suggests the renewables bloc is broadly-based, across government departments and corporate associations as well as a range of non-government organizations, climate campaigns and think tanks. This contrasts with the relatively narrow range of participants in the coal bloc, which centres on state enterprise, large corporates and relevant government departments. While this may reflect the logic of incumbency, Shukla and Swarnakar suggest it points to a dynamism in the renewables bloc that is likely to strengthen, offering a 'mere possibility' for addressing coal dependency (2023, p. 11). Pointers for future research are especially rich, 'to deploy more coherent ideological devices against the grand narratives supporting the fossil-fuel regime' (2023, p. 13).

With the capture of the liberal democratic state by fossil fuel interests, democratic climate change policy-making is clearly compromised at the national level. It is arguably even less evident at the international level within inter-state climate governance bodies such as the Intergovernmental Panel on Climate Change (IPCC), and within the UN system itself. This is a general problem for any civil society group addressing an inter-state body, but is particularly evident for Indigenous Peoples, where there can be a profound clash of knowledge systems. Inter-state agencies have sought to address their structural democratic deficit themselves with an array of ostensibly deliberative consultative forums. Through this, Indigenous Peoples have made significant advances in terms of demanding and acquiring consultative status, which itself has enabled substantive gains, including the ground-breaking UN Declaration on the Rights of Indigenous Peoples. In the sphere of inter-state climate policy-making, Indigenous Peoples have gained limited consultative status under the UN Framework Convention on Climate Change with the creation of a formal UNFCCC Facilitative Working Group on 'Local Communities and Indigenous Peoples' in 2018, with its own platform and workplan. Consultative status can enable recognition but it does not necessarily alter the way in which Indigenous issues are addressed.

Focusing on the IPCC, Rashidi and Lyons highlight 'colonial environmental governance' as the key structural barrier to Indigenous knowledges in climate change assessments. The IPCC is explicitly established as an inter-state agency designed to engage with the scientific 'community'. Yet it offers only limited recognition of Indigenous Peoples concerns. Within the IPCC, Rashidi and Lyons argue, climate science is presented as a scientific problem rather than a social-cultural problem. In this context, Indigenous perspectives are relegated as evidence, grist to the scientising process, rather than interpreted in their own right. Indigenous Knowledges are 'treated primarily as *sources* of data or *information* that can *supplement* science if they prove useful' (Italic in original; 2023, p. 4). This utilitarian approach reflects the assumption that 'climate change can only be imagined from the perspective of Western science' (2023, p. 7). More holistic Indigenous interpretations that may encompass wider relations with ecology, are marginalized, relegating Indigenous agency. To address this, Rashidi and Lyons suggest means of achieving a deeper dialogue, as an 'equal exchange' between Western science and Indigenous Knowledge over the implications of climate change (2023, p. 11).

Participation?

Energy democracy, in terms of public and community ownership of renewable energy, has become critical for climate justice, both in terms of energy autonomy and also in terms of offering a model for distributed emissions reduction. The dangers of reliance on the development of large-scale renewable energy installations, owned by corporates and government, are becoming clearer. At one level these installations deliver 'at scale' what is required in terms of the capacity to replace fossil fuel power generation; they also deliver 'at speed' in terms of being available for rapid deployment. But they are generally not socially replicable and leave the rest of society relatively untouched, maintaining the pre-existing reliance on large-scale utilities producing for energy 'end users'. The projects industrialize farmed, forested or so-called 'vacant' land, and can have major local impacts in terms of livelihood, both for those who own the land and may benefit from rental income and for those who do not own land and may lose informal access to lands and to agricultural employment. Clearly large-scale renewable energy has impacts that reflect local regimes in terms of land use, land ownership and stratification. Impacts are affected by the mode of technology that is deployed, for instance in terms of wind as against solar power, and the types of policy instruments that are used to establish the operations. These various factors then condition the possibilities for democratic engagement and involvement, which may change over time, reflecting strategies of the players, both at particular installations and from one installation to the next, as a process of experimentation and development unfolds.

These questions are clearly posed by Devleena Ghosh, Gareth Bryant and Priya Pillai in their account of one of Asia's largest solar farms in Pavagada in the State of Karnataka, Southern India, constructed in 2020. The scale of the solar farm, 12,000 hectares with a capacity of 2GW, far exceeds the size of any solar farm in for instance Germany or Australia. The speed at which it was constructed, from approval in 2015 to opening in 2019, is also exceptional. The lessons it holds are significant, not least as India and many other countries plan to 'scale-up' solar installations. In 2022 for instance, Karnataka was already planning a 'hybrid' wind and solar plant almost triple the size at 5GW, learning from the Pavagada experience.

Bryant et al. report on ethnographic engagement with local people affected by the project. Key questions centre on land use and land justice. The Karnataka and federal authorities established a joint agency to acquire land and establish infrastructure in what was identified as a drought-prone and relatively deprived region. Global solar operators competed for plots in a series of reverse auctions for multi-decade energy supply agreements. The 'special purpose' agency mediated between local landholders, private solar investors, and state-owned distributors. The agency leased the land, set a standard rental payment and persuaded local landholders to sign on. As Bryant et al. demonstrate, the model delivered very cheap electricity, but at the cost of local livelihood. The land is degraded and scarce water is directed to panel cleaning. Work in the solar plants is limited, and those without land, half of the local population, lose access to agricultural work and to grazing land. Stratification, as a result, is sharpened. Bryant et al. quote a landless woman worker who pithily summarizes: 'People who had more land become wealthier. We are where we are, landless people' (2023, p. 9).

Bryant et al. find these issues had been anticipated, notably by the World Bank, but that local people report there had been no meaningful action to address them. The case demonstrates the centrality of land ownership structures, along with other implications, for future utility-scale renewables; it also demonstrates the need for other options, especially for more distributed means of securing energy transition. The critical question is the overall approach taken. As Bryant et al. argue:

> when renewable energy policy is primarily concerned with reducing costs and managing risks for private investors and state actors, over a broader set of social, environmental, political, and economic goals, the potential benefits of energy transition will not be equitably shared (2023, pp. 2–3).

So-called 'least-cost abatement' invariably involves cost-shifting, to those least able to pay.

Similar issues emerge elsewhere, including in the radically different contexts of Germany and Australia. Franziska Mey and co-authors Anne Kallies, John Wiseman, and Michel Watson, select two high-income countries, Germany as a climate frontrunner and Australia as a climate laggard, and find parallel contexts and problems. For them, the question of democratic legitimacy for energy transitions is 'a prerequisite for successful energy transitions' (2023, p. 12). Renewable energy has strong general 'input' support in both countries but has faltered in terms of 'output legitimacy', in terms of how policy is implemented: as they note, 'energy transitions can create winners and losers in the community, with potentially severe impacts on the legitimacy, justice, and fairness of the energy transition' (2023, p. 13). They explore the role of public participation in energy transition, stressing the need for community ownership and the important framing aspects in 'instrument choice' for renewable energy policy (2023, p. 4).

Their analysis of the German energy transition points to the important shift from community and public ownership to more corporate-centred models from 2012. The policy shift was from a renewables feed-in tariff to a reverse auction model, 'an auction approach fostering only large-scale renewable energy developments to be conducted by large developers and utilities while systematically marginalizing small- and medium-sized community actors' (2023, p. 6). They track the rapid fall-off in community renewables, and the relegation of community participation to planning rather than owning, delimiting benefit. No surprise there had been an increase in public criticism of energy transition, especially in terms of the inequities of corporate as against community benefit. Australia presents similar tendencies, with strong growth in household rather than community or municipal renewables based on household feed-in tariffs at the sub-national State level. At the same time, there are large-scale corporate-led installations often centred on reverse auctions ('power-purchase agreements'). The impacts are felt in terms of increased contestations over planning decisions and growing public disquiet, especially in the few places with sufficient transmission capacity. As Mey et al. note, this is a product of policy: 'due to the design of support mechanisms, renewables investment in Australia clusters either on the large-scale or individualized household-scale ends of the system'; neither, as they note, is 'conducive to … community participation and buy-in' (2023, p. 11). The antidote is largely the same for both countries: stronger policy frameworks to enable community-level ownership of renewable energy, to thereby strengthen legitimacy for the transition.

Jonathan Marshall's article comparing community and municipal renewable energy projects extends the discussion of policy legitimacy. Echoing May et al, he shows how renewable energy policy is geared to large corporate interests, against more democratic community-level and municipal initiatives. The 'climate oligarchy', as he describes it (2023, p. 1), has a stranglehold on energy policy, constraining non-corporate renewables and (mis)directing innovation. He shows how local government councils in three small towns have struggled to introduce solar power. In all three cases, they have encountered regulatory restrictions on renewables and carbon lock-in, preventing them from creating community-level renewable projects. In contrast, corporate-owned renewables are favoured, with energy policy designed to capture 'end users' and realize a return for renewable energy companies. The pressure to capture consumption grows with declining demand: in NSW electricity consumption has been static for more than a decade and a growing proportion of this is met by household solar. There is no surprise that in this context more distributed community and municipal initiatives are not favoured. Marshall's ethnographic account draws out the contrasting models for community and municipal energy. He finds a spectrum of approaches and experiments, and identifies the particular case of solar gifting, what he calls 'climate generosity' (2023, p. 11), as the most effective. The 'gift economy' thrives on community initiative and public purpose, sidestepping the neoliberal 'business models' embedded in climate policy. With this it offers a counterpoint that invokes and expresses the scope and urgency of the climate challenge that it seeks to address. With this, as Marshall argues, promises much as a form of social action, offering more participatory and more democratic modes of energy.

Mobilization?

Katja Müller outlines the role of climate camps in Germany since 2011, which have been important in providing the infrastructure for large-scale protests away from cities and towns. They have become a focus for networking, education, organizing and action, and are themselves 'system-critical' laboratories of climate democracy. Regular camps have been organized at sites of coal extraction in the Lausitz region (Lusatia) in Eastern Germany and in the Central German mining area (Mitteldeutsches Revier). These have brought together local people opposing the destruction of their villages and surrounding landscapes, linking them with protesters from outside the coal regions who are concerned about coal and climate change. They have enabled a climate framing for local groups and, reciprocally, they have delivered a local grounding for climate activism. Some climate camps have provided the basis for disruptive direct action, such as the Lausitzcamp in 2016, at which protesters entered the nearby Welzow-Süd mine and occupied earth movers and railway tracks, interrupting operations at the mine and the adjacent power plant. The camps and associated protests have attracted significant media attention in Germany, and beyond, and have been an effective strategy for the climate movement to draw attention to coal's localized contribution to climate change.

Between 2017 and 2021, the mining and energy company LEAG, which owned both the mines and power plants in the Lausitz, announced it would discontinue plans to expand its mining operations in the region, stating that its decisions were purely economic. Over the same period, the German Federal Government set up the Commission on Growth, Structural Change and Employment ('Coal Commission') and legislated an exit from coal-fired power by 2038. As Müller concludes, while it cannot be clearly defined how much the climate camps and the climate movement in general contributed to these developments, 'the concurrent strengthening of the German climate movement and the development of fossil fuel critical national politics is striking' (2023, p. 5). Moreover, the camps themselves, with their key demand for 'system change not climate change', not only call for a reconsideration of political decision making, consumption processes and ownership structures but 'attempt to put those demands ... into practice' (2023, p. 2). While these aspirations may not always be realized, the camps enabled participants to imagine, test, and experience different democratic structures and processes and alternative ways of life.

Patrick Bond's contribution shows how much more difficult and dangerous it can be for activists in the global South to confront the coal industry and press democratic claims. Fikile Ntshangase was shot six times by three men who entered her house as her grandchildren were playing nearby. The motive for her murder is widely believed to have been her refusal to withdraw a court challenge to the expansion of the Tendele coal mine, close to Somkhele village where she lived, and bordering the Hluhluwe-iMfolozi Park, a vital site of conservation for the Southern White Rhinoceros in the 1960s. Ntshangase was taking action on behalf of local farmers who claimed the mine was contaminating their land and consuming vast quantities of river water they relied on for their livelihood. The mine is owned by Petmin, a Johannesburg-based company originally funded the by International Finance Corporation, a venture capital arm of the World Bank. Petmin claimed the Tendele mine was the primary source of employment for a community of 220,000 people and would have to close after two years unless the proposed expansion went ahead. Shortly before her death, Ntshangase had told friends she had been offered a US$24,000 bribe by mine management to drop her legal action. Ntshangase was supported by the MCEJO (Mfolozi Community Environmental Justice Organization), a group of allied organizations and lawyers with around 4000 members. Ntshangase and MCEJO found themselves in conflict with powerful mining unions, which backed the mine expansion, as well as with the company and its government supporters.

Bond argues that 'an exploration of the strengths and weaknesses of oppositional climate politics is one way to pay tribute to Ntshangase' (2023, p. 2). He argues that environmental justice groups must find ways to move beyond their clash of interests with mineworkers and their unions and explore the potentials for 'uniting with labour to demand and win a genuine Just Transition'. Without the alternative of a Just Transition, writes Bond, the mineworkers' unions were logically opposed to Ntshangase's fight to stop the mine expansion, and to broader environmental and climate considerations. He argues a key aspect would be the creation of a regional development fund to enable the community to disengage from fossil fuels, and instead generate new forms of livelihood. This, he suggests, could be funded via financial reparations from the company and its funders, including the World Bank, an approach already by pursued by MCEJO's lawyers. A wider strategy would entail 'a northern down-payment on climate debt', a to pay South Africa to refuse continued operations, perhaps on a global 'climate easement' model, to enable new forms of livelihood, for social and environmental regeneration in KwaZulu-Natal.

Climate movements have undergone a series of phases since the early 1990s, with each phase informing the next, in a cumulative process. The developments charted by Müller and Bond underline the centrality of on-the-ground mobilization in generating new linkages and new leverage. The critical strategic questions emerge and are answered organically, within the process of challenging and mobilizing against fossil fuel hegemony. James Goodman and Tom Morton show how the mobilization has become more, not less, central to the movement. In this, climate movements reverse the established social movement trajectory, from politicization to mobilization to institutionalization (and normalization). With the authority of climate science, climate movements have their origins in policy advocacy in the 1990s, as a form of 'climate action' advocacy centred on international negotiations and led by environmental non-government organizations. In the mid 2000s the failure of climate action forced new justice-centred climate issues onto the agenda, with a new wave of mobilizations specifically targeting fossil fuel culprits. A decade later, with the rapid acceleration of climate change, a third wave of mobilization has erupted centred on the impacts of climate disruption, and expressed most powerfully in the mobilization of outrage at the resulting intergenerational injustice.

Goodman and Morton focus on recent experience in India, Germany and Australia to discuss drivers and characteristics of these three waves. They find an intensification of mobilization: from climate petitions, earth hour actions and 'walks against warming', to mine blockades, climate camps and school strikes, and then more recent efforts at shutting the city, jamming culture, and dislocating 'normality'. Climate organizing has broadened across virtually every social field, and the climate agenda has deepened from emissions reduction to fossil fuel phase-out, to green 'new deals', and increasingly to 'de-growth'. With ongoing failure to reduce greenhouse gas emissions and intensifying climate change we can only expect the political challenge to grow. Certainly, as we have discussed in this introductory article, there are now more far-reaching measures on the agenda, both in terms of movement demands and in terms of movement actions.

Conclusion

Climate change poses a profound challenge for liberal democracy. If, as we have argued, indirect representative democracy is unable to shed itself of the power and influence of the fossil fuel sector, then it is also unlikely to be able to protect itself and wider society from the ravages of climate disruption. Already it is clear that formal liberal democratic politics is unable to respond effectively to climate change. The question this poses for other forms of democratic involvement, including for mobilization and civil disobedience, is central. The debate about organizing approaches, both inside and outside the political 'tent', suggests the power of connective and generative action, for new claims as well as new forms of action. The debates about failed policy, and the need for alternatives, point to more thorough-going measures that can address the causes of climate instability 'at source'. These climate claims and linked climate actions are repositioned as critical dimensions of democratic life, mobilizing new constituencies that can overwhelm and ultimately transform existing representative systems.

Within these dynamics, the crucial institutional questions of state power remain underdeveloped. There is a tendency for the movement to find or claim democratic spaces above and below the national state. There are initiatives for inter-state fossil fuel moratoriums and non-proliferation, for payment of climate debts or reparations, and transfers to maintain carbon sinks or prevent fossil fuel extraction. At the local level movements establish 'intentional' communities of deliberation and action and strengthen local democratic capacity and legitimacy, in a 'pluriverse' of alternative, ecologically-engaged structures (Kothari et al., 2019). Notwithstanding these inter-state and local-level flanking efforts, national states continue to exercise sovereignty for fossil fuels. This is a deliberate exercise of power: options exist and choices are being made between more or less climate-destructive roles for the state. The ubiquitous climate league tables are part of the effort to put downward pressure on fossil fuels, along with narratives of decarbonized prosperity. Capacity to umask the rhetoric, and expose co-option via 'transformismo', has become critical (see Gills & Morgan, 2021). The challenge is daunting: geopolitical dynamics, with the state and its vested interests pitted in world-systemic struggles for advancement are posed against the symbolic and deliberative power of global climate mobilizations. Here, the inadequacy of the elite's 'passive revolution' and the growing salience of climate disruption force a new dynamism into the process.

That dynamism generates traction for new more far-reaching climate action agendas. As we have argued, following O'Connor, the climate contradiction at the heart of prevailing capitalist society requires agendas for socialized provision (O'Connor, 1998). The continued failure to shift state policy in the face of growing climate crisis only forces more transformative approaches onto the table. More interventionist state roles are emerging both to decarbonize and socialize provision, as expressed in the 'new deals' debates. The limits of these themselves point to wider agendas for post-growth, and for post-capitalism. Indeed, the demand for 'system change not climate change' – to quote a slogan popular at climate protests around the world – has become a central point of debate and divergence within the climate movement itself, with some arguing that the urgency of rapid and far-reaching emissions reduction means that system change must take second place to decarbonization by whatever means possible, while others argue that decarbonization of our economies and societies will only be possible through a complete dismantling of the global capitalist system and its replacement by 'de-growth communism' (Saito, 2023). This tension, between reformist and revolutionary tendencies within the climate movement, between 'Green new deals' based in ecological modernization and more far-reaching democratization of the ecological commons and the modes of production and consumption of energy, is inextricably linked to questions of justice; and demands for climate justice are increasingly centred on, and driven by, the participation and mobilization of Indigenous Peoples, and social movements in the Global South (Kröger, 2020), including in struggles against fossil fuel extraction and for stronger climate action.

As we stressed at the outset, questions about how to overcome the stranglehold of carbon democracy, and define new possibilities for climate democracy, are being forged in the process of challenging the fossil fuel bloc. Within this, the process of praxis and reflection to generate new agendas for action, is critical. Historical perspective on how gains have been made, across strategies of climate action, climate justice, and climate emergency, becomes vitally important. Key mobilizing concepts and agendas – unburnable carbon, just transition, energy democracy, green new deals, de-growth, living well, the pluriverse, and Indigenous climate justice – emerge from the dialectical process of challenging carbon democracy. Reading these as aspects in an intensifying struggle offers a means of writing this history into the present, to chart future agendas. We hope this Special Issue contributes in some way to advancing these agendas.

Disclosure statement

No potential conflict of interest was reported by the author(s).

Funding

This work was supported by Australian Research Council [grant number DP180103868 Decarbonising Electricity].

ORCID

James Goodman http://orcid.org/0000-0003-3211-3416

References

Abourahme, N., & Jabary-Salamanca, O. (2016). Thinking against the sovereignty of the concept. *City*, *20*(5), 737–754. https://doi.org/10.1080/13604813.2016.1224486

Albert, M. (2021). The climate crisis, renewable energy, and the changing landscape of global energy politics. *Alternatives: Global, Local, Political*, *46*(3), 89–98.

Almeida, P. (2019). Climate justice and sustained transnational mobilization. *Globalizations*, *16*(7), 973–979. https://doi.org/10.1080/14747731.2019.1651518

Aranof, K., Battistoni, A., Cohen, D., & Riafrancos, T. (2019). *A planet to win: Why we need a green new deal*. Verso.

Arrighi, H., & Wallerstein, I. (1989). *Anti-systemic movements*. Verso.

Aykut, S., Foyer, J., & Morena, E.2019). *Globalising the climate: COP21 and the climatisation of global debates*. Routledge.

Beck, U. (2015). Emancipatory catastrophism: What does it mean to climate change and risk society? *Current Sociology*, *63*(1), 75–88. https://doi.org/10.1177/0011392114559951

Böhm, S., & Sullivan, S. (2022). *Negotiating climate change in crisis*. Open Book Publishers.

Burke, M., & Stephens, J. (2017). Energy democracy: Goals and policy instruments for sociotechnical transitions. *Energy Research & Social Science*, *33*, 35–48. https://doi.org/10.1016/j.erss.2017.09.024

Burnell, P. (2012). Democracy, democratization and climate change: Complex relationships. *Democratization*, *19*(5), 813–842. https://doi.org/10.1080/13510347.2012.709684

Carney, M. (2015). *Breaking the tragedy of the horizon – climate change and financial stability*. Gov BoE.

Carroll, W. (2020). Fossil capitalism, climate capitalism, energy democracy: The struggle for hegemony in an era of climate crisis. *Socialist Studies*, *14*(1), 1–26.

Chagnon, C., Durante, F., Gills, B., Hagolani-Albov, S., & Hokkanen, S. (2022). From extractivism to global extractivism: The evolution of an organizing concept. *The Journal of Peasant Studies*, *49*(4), 760–792. https://doi.org/10.1080/14747731.2022.2142013

Chakrabarty, D. (2021). *The climate of history in a planetary Age*. Chicago University Press.

Connolly, K. (2023). German Police stage nationwide raids against climate activists. *The Guardian*, 24 May.

Crutzen, P. J., & Stoermer, E. (2000). The anthropocene. *Global Change Newsletter*, *41*(1), 17–18.

Darian-Smith, E. (2022). *Global Burning: Rising anti-democracy and the climate crisis*. Stanford University Press.

Die Zeit. (2022). Pistorius: Vergleich von Aktivisten mit RAF ist "absurd", 14 November.

Editorial. (2020). In the line of fire. *Nature Climate Change*, *10*(3), 169–169. https://doi.org/10.1038/s41558-020-0720-5

Farand, C. (2022). Saudi Arabia dilutes fossil fuel phase out language with techno fixes in IPCC report. *Climate Home News*, 4 April 2022.

Fioirino, D. (2018). *Can democracy handle climate change?* Polity.

Gills, B., & Frank, A. (1994). *The world system five hundred years or five thousand?* Routledge.

Gills, B., & Morgan, J. (2021). Postscript, an end to the war on nature: COP in or COP out? *Globalizations*, *18*(7), 1311–1322. https://doi.org/10.1080/14747731.2021.1973273

Goldhaber, D. (2023). How to blow up a pipeline. *Feature Film*, Neon, New York.

Goodman, J., (2021). Three worlds of climate imperialism: Prospects for climate justice'. In H. Hosseini (Ed.), *The Routledge handbook of transformative global studies* (pp. 301–314). Routledge.

Goodman, J., Ghosh, D., Connor, L., & Kohli, K. (2020). *Beyond the coal rush: Turning point for global energy and climate policy?* Cambridge University Press.

Goodman, J., & Morton, T. (2014). Climate crisis and the limits of liberal democracy? Germany, Australia and India compared. In B. Isakhan, & S. Slaughter (Eds.), *Crisis and democracy in the twenty-first century: Democratizing Governance* (pp. 229–252). Palgrave.

Guterres, A. (2022). *Secretary-General's video message to the Press Conference Launch of IPCC Report*. United Nations Secretary-General. 22 February.

Hafner, M., & Tagliapietra, S. (2020). *The geopolitics of the global energy transition*. Springer.

Haunss, S., Rucht, D., Sommer, M., & Zajak, S. (2019). Germany, in Wahlström, M. et al, protest for a future: Composition, mobilization and motives of the participants in Fridays for future climate protests on 15 March, 2019 in 13 European cities, 67–80.

Head, L. (2020). Transformative change requires resisting a new normal. *Nature Climate Change*, *10*(3), 173–174. https://doi.org/10.1038/s41558-020-0712-5

Hess, D. (2018). Energy democracy and social movements: A multi-coalition perspective on the politics of sustainability transitions. *Energy Research & Social Science*, *40*, 177–189. https://doi.org/10.1016/j.erss.2018.01.003

Huber, R., Greussing, E., & Eberl, J. (2022). From populism to climate scepticism: The role of institutional trust and attitudes towards science. *Environmental Politics*, *31*(7), 1115–1138. https://doi.org/10.1080/09644016.2021.1978200

Hulme, M. (2010). Mapping climate change knowledge: An editorial essay. *WIRES Climate Change*, *1*(1), 1–8. https://doi.org/10.1002/wcc.3

IEA. (2021). *World energy outlook: Executive summary*. IEA.

Influence Map. (2022). *Corporate climate policy footprint: The 25 most influential companies blocking climate policy globally*. Influence Map, London.

IRENA. (2020). *Global renewables outlook: Energy transformation 2050*. International Renewable Energy Agency.

Kallis, G. (2020). *The case for degrowth*. Polity Press.

Kothari, A., Salleh, A., Escobar, A., Demaria, F., & Acosta, A. (2019). *Pluriverse: A post-development dictionary*. Tulika Books.

Kröger, M. (2020). *Iron will: Global extractivisms and mining resistance in Brazil and India*. University of Michigan Press.

Kurmelovs, R. (2022). *The sentencing of climate activist Violet CoCo. The Saturday Paper*. 10 December.

Kusnetz, N. (2021). Driven by industry, more states are passing tough laws aimed at pipeline protesters. *Inside Climate News*, 12 April.

LeQuesne, T. (2020). From carbon democracy to carbon rebellion: Countering petro-hegemony on the frontlines of climate justice. *Journal of World-Systems Research*, *25*(1), 15–27. https://doi.org/10.5195/jwsr.2019.905

Macalister, T. (2022). UK climate activists held in jail for up to six months before trial. *The Guardian*, 23 September.

Malm, A. (2021). *How to blow up a pipeline. Learning to fight in a world on fire*. Verso.

McCarthy, J. (2015). A socioecological fix to capitalist crisis and climate change? The possibilities and limits of renewable energy. *Environment and Planning A: Economy and Space*, *47*(12), 2485–2502. https://doi.org/10.1177/0308518X15602491

McCauley, D., & Heffron, R. (2018). Just transition: Integrating climate, energy and environmental justice. *Energy Policy*, *119*, 1–7. https://doi.org/10.1016/j.enpol.2018.04.014
Milman, O. (2022). *What's in the climate bill that Joe Manchin supports - and what isn't. The Guardian*. 29 July.
Mitchell, T. (2009). Carbon democracy. *Economy and Society*, *38*(3), 399–432. https://doi.org/10.1080/03085140903020598
Mitchell, T. (2013). *Carbon democracy: Political power in the age of oil* (3rd ed.). Verso.
Mookerjea, J. (2019). Renewable energy transition under multiple colonialisms: Passive revolution, fascism redux and utopian praxes. *Cultural Studies*, *33*(3), 570–593. https://doi.org/10.1080/09502386.2019.1585464
Moore, J. (2015). *Capitalism in the web of life: Ecology and the accumulation of capital*. Verso.
O'Connor, J. (1998). *Natural causes: Essays in ecological marxism*. Guilford Press.
Ollinaho, O. (2023). What is 'business as usual'? Towards a theory of cumulative sociomaterial change. *Globalizations*, *20*(4), 611–627.
Parkes-Hupton, H. (2022). NSW parliament passes new laws bringing harsher penalties on protesters. *ABC News*, 1 April.
Prytherch, M., Lieberthal, K., & Hass, R. (2023). *Unpacking China's climate change priorities*. Brookings Institution.
Richards, L., Brew, N., & Smith, L. (2020). *2019–20 Australian bushfires: A quick guide*. Parliamentary Library, Commonwealth of Australia. 12 March.
Rose, T., & McGowan, M. (2022). NSW premier describes jailing of climate activist Deanna 'Violet' Coco as 'pleasing to see'. *The Guardian*, 3 December.
Rosemberg, A. (2010). Building a just transition: The linkages between climate change and employment. *International Journal of Labour Research*, *2*(2), 125–161.
Saito, K. (2023). *Marx in the anthropocene: Towards the idea of degrowth communism*. Cambridge University Press.
Scheidel, A., Del Bene, D., Liu, J., Navas, G., Mingorría, S., Demaria, F., Avila, S., Roy, B., Ertör, I., Temper, L., & Martínez-Alier, J. (2020). Environmental conflicts and defenders: A global overview. *Global Environmental Change*, *63*, 102104. https://doi.org/10.1016/j.gloenvcha.2020.102104
Shearman, D., & Smith, J. (2007). *The climate change challenge and the failure of democracy*. Bloomsbury.
Shih, G., Mehrotra, K., & Gupta, A. (2023). India cracks down on critics of coal. *Washington Post*, 5 June.
Steger, M., Goodman, J., & Wilson, E. (2013). *Justice globalism: Ideology, crises, policy*. Sage.
Szulecki, K. (2018). Conceptualizing energy democracy. *Environmental Politics*, *27*(1), 21–41. https://doi.org/10.1080/09644016.2017.1387294
Tigue, K. (2022). Bold climate protests are triggering even bolder anti-protest laws. *Inside Climate News*, 22 November.
Urry, J. (2014). The problem of energy. *Theory, Culture and Society*, *31*(5), 3–20. https://doi.org/10.1177/0263276414536747
US Congress. (2019). *Recognizing the duty of the Federal Government to create a Green New Deal*. House Resolution 109, Congress of the United States.
Van Veelen, B. (2018). Negotiating energy democracy in practice: Governance processes in community energy projects. *Environmental Politics*, *27*(4), 644–665. https://doi.org/10.1080/09644016.2018.1427824
Wainwright, J., & Mann, G. (2013). Climate leviathan. *Antipode*, *45*(1), 1–22. https://doi.org/10.1111/j.1467-8330.2012.01018.x
WWF. (2020). *Australia's 2019–2020 bushfires: The wildlife toll*. World Wildlife Fund.

Special Issue Papers

Bond, P. (2023). Climate, violence, resource extraction and ecological debt: Global implications of an assassination on South Africa's coal mining belt. *Globalizations.*

Bryant, G., Ghosh, D., & Pillai, P. (2023). Who wins and who loses from renewable energy transition? Large-scale solar, land and livelihood in Karnataka, India. *Globalizations.*

Goodman, J., & Morton, T. (2023). Climate movements in Germany, India and Australia: Dynamics of transition, transformation and emergency. *Globalizations.*

Marshall, J. (2023). Comparing local energy conflicts in NSW Australia: Moving to climate generosity. *Globalizations.*

Mey, F., Kallies, A., Wiseman, J., & Watson, M. (2023). Legitimizing energy transitions through community participation: Germany and Australia at a crossroads. *Globalizations.*

Müller, K. (2023). Climate camps and environmental movements. Impacting the coal industry and practicing 'system change'. *Globalizations.*

Plehwe, D. (2023). Reluctant transformers or reconsidering opposition to climate change mitigation? German think tanks between environmentalism and neoliberalism. *Globalizations.*

Rashidi, P., & Lyons, K. (2023). Democratizing global climate governance? The case of indigenous representation in the Intergovernmental Panel on Climate Change (IPCC). *Globalizations.*

Shukla, R., & Swarnakar, P. (2023). Energy transition and dialectics: Tracing discursive resistance to coal through discourse coalition in India. *Globalizations.*

Part I

Representation

OPEN ACCESS

Reluctant transformers or reconsidering opposition to climate change mitigation? German think tanks between environmentalism and neoliberalism

Dieter Plehwe

ABSTRACT

Germany is known as a leader in energy transition, but the country keeps falling short of meeting stated objectives. While attention has been high on climate change related policy leadership, the competing domestic alliances in the field of energy transition have received scant attention. In Germany, the strong momentum of the conversion to renewable energy has been stalled in the second half of the 2010s. Enabled by an increasing weight of neoliberal economic concerns, traditional energy coalitions have managed to reshape the path of energy transition. In this article three competing discourse coalitions are introduced to examine their constituencies and projects in the field of energy and politics. It focuses on radical ecological forces around the Öko-Institut, environmental modernization capacities linked to Green Budget Germany and fossil interest groups intersecting with neoliberal academic and partisan think tanks. As such, the article opens a view on multiple contestations over Germany's Energiewende.

1. Introduction

Germany is generally considered a progressive force in climate change politics. Angela Merkel promoted herself as a global leader in this field much like the preceding Christian Democratic government of Helmut Kohl back in the 1980s and 1990s, let alone the Social-Democrat-Green Coalitions headed by Gerhard Schröder (1998–2005). Supported by a spearheading role in important areas of climate change mitigation – emission reduction targets in the 1990s, expansion of renewable energy in the 2000s – the country attracted the home office of the United Nations Framework Convention on Climate Change in Bonn in 1994 and of the Innovation and Technology Centre of the International Renewable Energy Agency (HQ Abu Dhabi), which was also founded in Bonn in 2009. Germany is the third largest producer of renewable energy excluding hydropower (solar and wind, mainly) following the much larger countries U.S. and China, and has a much larger per capita share of renewables than the U.S and China. In terms of per capita share, Germany is trailing the much smaller nations Iceland, Denmark and Sweden only (Renewables, 2021, p. 41).

But Germany's climate change policy making has not been without fault and blame. As pointed out by several authors, Germany's practical commitments kept falling short of stated objectives. Ambitious greenhouse gas reduction goals have been subject to backsliding and political horse trading to protect strong industrial interests both in the national arena and in the European Union. Policy making has been marred by divisions between various responsible ministries and the federal system of power sharing between federal and state level (Länder) governments. Last but not least, corporatist traditions have enabled influence peddling by incumbent energy firms and industrial interest groups in particular (Michaelowa, 2008, 2013; Stevenson, 2021).

The mixed record of German climate change mitigation has led one expert to characterize Germany as a 'climate policy giant on earthen feet' (Michaelowa, 2013, p. 319). Germany can thus be considered a perfect site for a case study both on 'path dependencies' and 'institutional change' in the area of climate change mitigation. Due to the zig-zag policy around nuclear phase out[1] and dedicated support for renewables, Germany is also a good place to look for 'critical junctures' and 'policy change' in the field of energy policy in particular (Stefes, 2014).

In spite of strong claims to climate leadership and renewable energy conversion, Germany surprisingly is also a perfect place to study the politics of climate policy opposition and obstruction, i.e. strategies developed and executed to prevent ambitious mitigation, to slow down climate or to reverse policy developments considered detrimental to fossil energy and other corporate interests (Brulle & Aronczyk, 2019; Lamb et al., 2020). In order to observe conservative and reactionary efforts to shape climate change related policy making, a broad range of both academic and partisan think tanks as well as NGOs and campaign organizations need to be included in the analysis in addition to traditional (fossil) interest and corporate lobby groups. Michaelowa (2008) has shed some light on disruptive strategies in German climate policy making, though his focus has been on concentrated power of both industrial interests and renewable energy producers in their parts of the energy sector. In contrast to such an approach it is important to discuss the relationships between a wider range of actors and strategies involved in the continuing effort to undermine emission reduction targets and to shape mitigation and adaptation strategies. While solar groups were temporarily successful indeed reaping extraordinary profits from the pricing mechanism introduced in 2000, fossil groups succeeded to remove the feed-in tariff as policy instrument of choice, and in blocking the rise of renewables in the second half of the 2010s. In order to properly examine the relationship between industrial policy, competitiveness concerns and the energy transformation path chosen, it is important to account for a broad range of national and European, political and discursive strategies of the climate change policy opposition rather than focusing on market power in individual segments of the energy system.

Beyond the opposition to rapid decarbonization and energy transformation narrowly conceived, the business minded coalitions face the more fundamental challenge of the 'energy democracy' movement. This ecological 'discourse coalition' (Hajer, 1993) has been fundamentally critical to corporate and financialized capitalism and the consumerist growth regime. The coalition rose during the fight against nuclear energy production to advance an alternative economic approach in which the potential of decentralized renewables plays a major role (Morris & Jungjohann, 2016). In between the two camps that are diametrically opposed – corporate fossil conservatism versus overlapping movements of energy democracy, radical ecology, climate justice and 'degrowth' – we can trace the emergence of a new version of economic modernization under the label of an 'ecological social market economy'. This cross-party coalition plays a mediating role in Germany and elsewhere. With climate denialism thoroughly discredited, climate change policy obstruction needs to strategically advocate for a position between status quo protection and reformism (greener capitalism) in order to keep the more fundamental opposition to fossil capitalism at bay, and to secure longer transition periods. The political struggles over the transformation pathway are also struggles over the orientation and meaning of the sub-discipline of environmental economics as we shall see.

In order to situate contemporary struggles in the German climate and energy political field it is necessary to link energy conflicts to broader dimensions of international political economy, social geography, material infrastructures and power relations (Moss & Gailing, 2016). Geels (2014) has rightly called for greater attention to politics and power in the often technical and innovation-minded debate on energy transition. It is crucial to develop a conflict theoretical focus on policy related knowledge, ideas and discourse, because agents and agencies need to cognitively relate to each other to interpret interest positions, and because *discourse coalitions* influence policy and discourse in ways that work beyond economic interest positions narrowly conceived. Political and intellectual entrepreneurship in conjunction with particular interests group dynamics is not only relevant to explain institutional change and political dynamics (Becker, Beveridge, & Röhring, 2016), but also to observe resistance to reforms and policy obstruction (Bonds, 2011).

The contemporary opposition to transformative change needs to be creative in the face of new and strong challenges to incumbent positions and dwindling support for the status quo in the face of climate emergencies. Possibly because of the greater relevance of specialized knowledge and the volatility of many policy fields, the supply of policy-related expertise has been increasingly externalized by economic interest groups. A wide range of policy actors can be observed employing dedicated consulting firms and think tanks, sometimes as well as in-house experts, to strategically advance their interests. The growing number and diversity of academic and partisan think tanks in particular that are involved in struggles over energy transition and the climate change-related transformation process, exemplifies this trend.

I focus on think tanks and related campaign organizations both in terms of organizations and functions dedicated to policy related research, consulting and publicity. While focusing on contested policy expertise it is important to recognize that neither think tanks nor academic research institutes exist in isolation of a range of (political, economic, academic etc.) interests that come into play in policy related research in one way or another and need to be taken into account (Medvetz, 2012; Stone, 2007). Academic independence and quality can therefore be best enhanced by way of consideration and recognition of dependencies and biases of policy knowledge in addition to evidence base, academic rigour or methodological prowess. Think Tanks (and related organizations/networks) in any case are in the following analyzed according to bottom up 'logics of constituencies' (academic, partisan orientations, funders etc.) and top down or sideway 'logics of influence' (political expectations, contracting customers, competing forces etc.) in adaptation of Schmitter and Streeck's (1999) approach to study interest groups. While interest groups have 'membership logics', few think tanks have a relevant membership base like the Öko-Institut or Green Budget Germany discussed below. But all think tanks do have constituencies like funders and supporters, public and/or private, staff members etc. involved in relevant networks. Much like interest groups, think tanks operate in a social space subject to external pressure enabling or disabling research projects and selectivity in terms of approaches and orientations, and to pressures emanating from competing alliances. Both academic and partisan think tanks are therefore ideal to study both key concepts and relevant actor relations that eventually come to define larger discourse coalitions (compare Plehwe, 2015). The dynamics between such groups and coalitions is critical for policy in particular in circumstances of strong contestations. We will therefore turn our attention to them in the effort to better understand and explain the ambiguities of the 'Energiewende' in Germany (and the European Union).

We need to first examine the rise of the anti-nuclear and broader environmental movement in Germany. One of the important organizations created in militant protest was the Öko-Institut in Freiburg, which published the original Energiewende proposals. The institute combined outsiders from the mainstream Social Democratic Party with radical social movement forces together constituting the radical wing of the ecological social movement (Section 2). We need to secondly examine new efforts to advance agendas for ecological economic modernization and environmental economics, which subsequently gained traction in the German policy field. The centrist reconfiguration of the Green Party during the 1980s enabled the founding of the bipartisan Forum for the Ecological Social Market Economy, which advanced a model Green Budget for Germany in the 1990s (Section 3). Last but not least it is necessary to examine the even more recent efforts to fight back against Energiewende and renewable energy in particular. This struggle has been advanced with the help of dedicated energy experts and neoliberal economists. Their perspectives have been disseminated widely through academic and partisan think tanks and related (lobby) networks (Section 4). The current stop and go conundrum in solar and wind energy has been a result of the dynamic interplay of forces within this triangle of energy political discourse coalitions and their underlying economic and social constituencies within the broader context of the emerging European 'Green Deal'. The confluence of fossil interests (both electricity producers and large energy customers), competitiveness concerns and the influence of neoliberal (energy) economists has been able to stall the advance of renewable energy transition in the one area in which Germany had excelled in the new millennium: electricity generated from renewable sources (wind and solar in particular).

2. Green policy think tanks: the Öko Institute and the origins of the Energiewende

The emergence of extraordinarily novel, far-reaching, comprehensive and radical policy ideas under the Energiewende is surprising and needs explaining in the face of path dependencies, vested interests and bureaucratic inertia (Rüb, 2014). Explanations that rely primarily on moments of crisis, external shock and windows of opportunity (such as presented by the Fukushima nuclear accident in 2011 in the energy political field) cannot explain the origins and the development of alternative ideas, concepts and necessary expertise needed to create a new path and to promote a new policy development. The backdrop of Germany's energy transition has much to do with the post-war energy system 'status quo' *before* the battles of the 1970s, of relying mainly on domestic coal-based electricity generation and regional (regulated) private distribution monopolies that were closely intertwined with both public and private producers and public municipal services. The centralized energy monopoly structure had been reinforced by the increasing reliance on nuclear energy, which also led the electricity debate far beyond energy concerns to include questions of international and domestic security, military infrastructures and public health and safety concerns. The post-war energy system was stable due to multi-partisan political support backed by business groups and labour unions. But the energy coalition ran into deep trouble in the course of the 1970s when a rapidly growing social movement started to attack nuclear power and pollution from coal-fired plants. The ensuing struggles against nuclear energy and other hazards related to energy production and consumption led to new conceptual developments in the field of ecology (Hirschl & Vogelpohl, 2019), and to new contention over energy policy.

Regarding environmental politics, progressive think tanks were highly relevant in (a) (re-)developing an ecological understanding of the world after the Second World War due to grave consequences of rapid industrialization, mass production and consumption under Fordism, in (b) promoting the new environmental movement and in (c) configuring the emerging environment and climate policy field since the 1970s. Surprisingly, existing histories of environmental politics in Germany and the European Union have nevertheless been silent about the role of think tanks (Böcher & Töller, 2011; Meyer, 2011).

Contrary to the establishment think tanks like Resources for the Future (RFF) in Washington, DC, or Barbara Ward's International Institute for Environment and Development (IIED) in London, which grew out of the political mainstream (Strong, 2000, pp. 252–259; Engfeldt, 2009, p. 50), Germany's first environmental think tank was set up in the midst of intense social struggle. The Öko-Institut in Freiburg much like major environmental and peace movement organizations such as Friends of the Earth or Greenpeace was born during environmental policy battles of the 1970s. Most prominent was the fight against nuclear bomb testing and nuclear power plant construction involving both the development of new political strategies like civil disobedience and a new politics of counter-expertise. The former became the mainstay of social activism of organizations like Greenpeace, Friends of the Earth or Robin Wood. The politics of counter-expertise led directly to the establishment of green/environmental (movement) think tanks (McCormick, 1989).

Germany's Öko-Institut in Freiburg was set up to support activism against the construction of a nuclear power plant in Whyl on the French-German border (Rucht, 1980). Activists needed expertise to support their case in the courtroom and critical scholars speaking out against nuclear energy needed a supportive institution for their academic work because they were subject to marginalization in the universities. To this day, the Öko-Institut in Freiburg is one of the few think tanks with a strong membership base due to its roots in the social movement, which provides a degree of independence from the funding agencies and contract research on which it mainly relies. The line between the two types of civil society organizations, the NGO and Think Tank, in fact are blurred with NGOs engaging frequently in investigative research and publishing findings in reports not so different from movement think tanks. Vice versa, the role of ecologically partisan think tanks in support of movement activists resembles many of the NGO activities (Spangenberg, 2012). But the two are distinct in terms of perspective: the growth of the organizational format of the think tank can be explained by the specific requirements of short, medium- and longer-term research efforts related to policy making compared to the quicker and generally media and policy attention-oriented pace of campaigning.

Due to the focus on research and opportunities for both policy-related and academic studies, think tanks gained a particular role in developing and shaping important conceptual frameworks that can be found guiding (environmental) policymaking to this day (Smith, 1991; Fischer, 1993). Regardless of the inevitably political character of knowledge, research and science conducted in think tanks has been capable of mustering greater recognition and legitimacy than the work conducted by other advocacy organizations. Independent from the more or less legitimate truth claims (scientific reliability, evidence-based research and 'objectivity'), think tanks and researchers working in such organizations acquire a role of intellectual and conceptual entrepreneurship beyond immediate research and consulting services to specific causes. Think tanks have become critical organizations for the development of popular short cut explanations of troublesome phenomena (e.g. trees dying due to acid rain), or *story lines* (Saloma, 1984).

The founding statement of the Öko-Institut in 1977 relates the dramatic consequences of the ongoing destruction of the environment and the failure of the state in this regard, but it also emphasized the positive role of the judiciary in Germany. The institute wanted to establish an independent and critical research unit to offer a decisive instrument at the service of self-responsible citizens (Roose, 2002). Key constituencies of the Öko-Institut included value-conservative and to a certain degree industry-critical Social Democrats like Erhard Eppler and the centrist opposition to nuclear energy, many of them church going people. Among the activists a broad coalition ranged from moderate and even conservative environmentalists to members of Germany's marginal Communist party, and activists in the local, German and transnational struggle against nuclear power plants. Last but not least a small community of critical academics needs to be mentioned that looked for a more favourable environment for their research (Roose, 2002). Institutional logics relevant to the work of the Öko-Institut instead refer to top down and sideway influences on the organization, among which the courts and the domestic security apparatus originally played a major role before they were superseded by German and European ministries and other contractors offering ever increasing funds for dedicated environmental policy research.

In addition to the early research related to the fight against nuclear energy, the Öko-Institut established work lines related to chemicals, genetics, material flow (waste management and recycling), transport and environmental law. In addition to the original location in Freiburg, offices have been opened in Darmstadt (near Frankfurt am Main in Hessen) and later in Berlin. Partly shedding its ecological radicalism, the Öko-Institut has become a major contract research organization for public and private contractors. Originally highly critical towards the private sector in particular, the institute now considers itself a partner of industry and government rather than an element of the radical opposition (Scheytt, 2007). This is partly due to the mutual effort of reconciliation between the environmental movement and corporate management interests (Weidner, 1996). Although the Öko-Institut was certainly a part of the progressive wing of the fledgling Green party movement of the late 1970s and early 1980s, many of the agendas it helped to promote moved quickly from the margins to the mainstream of the political and managerial class. Generating key concepts for a new political era driven by growing concerns over global warming in the course of the 1980s, species extinction and related problems of new inequalities, mass migration and war contributed strongly to the mainstreaming of Öko-Institut content. Becoming part of the new mainstream, however, did not mean becoming uncontroversial.

Turning to the field of energy, it was the Öko-Institut that came up with the visionary concept of 'Energiewende' (Krause, Bossel, & Müller-Reißmann, 1980). The authors were first anticipating a decoupling of economic development and energy consumption. The Energiewende book relied to a certain degree on the work of the American physicist Amory B. Lovins who developed ideas for a turn towards soft energy supply in the United States (Lovins, 1976/1977). The German authors deconstructed the nuclear energy narrative according to which there was no alternative to oil and uranium to close the energy gap and to sustain growth and employment of an industrialized nation like Germany. To the contrary, the book showed the limited supply of both uranium and oil ('peak oil') that would necessitate alternative energy sources regardless of the other good reasons to give up nuclear energy in particular. Interestingly, the book did not yet develop a full anti-fossil energy concept and projected a continuing reliance on coal, which still was a significant domestic economic sector in Germany at the time. Both the growing concern with global warming and the limited relevance of German coal mining quickly led to the modification of the original perspective. The much more damaging impact of East German lignite added additional urgency to coal phase-out after unification, although local populations mobilize resistance in the affected regions, including forces from the far-right party Alternative für Deutschland (Morton & Müller, 2016). Historical modifications notwithstanding, 'Energiewende' was the first daring proposal to significantly shift Germany's energy system and to advance a model dedicated to comprehensive policy change including systematic support for renewable energy and public policy planning instruments such as subsidies for energy conversion and prize guarantees (feed in tariffs). The energy experts of the Öko-Institut in this way also ushered in a new era of ecological economics.

The rise of radical ecological think tank and NGO networks in the 1970s in any case led to a lasting transformation of the political party spectrum of Germany, supporting and sustaining the rise to the sprawling Green Party movement. The integration of the Greens into the political system in turn had a strong influence on the emergence of a second category, of mainstream environmental think tank.

3. The rise of environmental policy think tanks: ecological tax reform and the advance of (neoliberal) environmental economics

Compared to the bottom up logics of civil resistance behind early radical ecological think tanks and social movement NGOs, a second group of environmental think tanks have been founded already with an eye to playing an important part in the institutional evolution of the environmental policy field at national and international levels. The critical role of the multi-country Institute for European Environmental Politics (IEEP) is a good example. The organization was founded by the German refugee Konrad von Moltke, son of one of the officers killed by the Nazis after the failed attempt on Hitler's life in 1944.[2] Konrad von Moltke returned to Germany via the Netherlands, and originally helped from there to build European and transatlantic research and communication channels needed to develop the European and global (United Nations-centred) environmental policy field (backed by the Dutch royal Prince Bernhard and the European Cultural Foundation). Following the IEEP, the Wuppertal Institute and the Ecologic Institute in Berlin have continued the national, pan-European and transatlantic effort envisioned by Konrad von Moltke with networks stretching from Germany to Brussels and Washington, DC. In the UK, the New Economics Foundation played a similar role, in Scandinavia the Stockholm Environment Institute, and so forth. The ecological movements in the streets were complemented by a new transnational environmental policy elite (Plehwe, 2012).

The proliferation of mainstream environmental research institutes followed the institutionalization of the policy field from the late 1970s onwards. Contrary to established policy fields, environmental policy research started outside of universities and government research facilities in many countries. Both the interdisciplinary character and the private civil society-based roots set a precedent for the subsequent outsourcing of government research and the rise of public-private governance in research and policy making more generally, which of course became a hallmark of the neoliberal age.[3] Environmental studies in this way was novel both with regard to the interdisciplinary character and with regard to the organizational format of the private European policy think tank contracting from government agencies at national and international levels.

Policy-related think tank research functions at the same time do not necessarily require a single organizational shell. Consider the crucial (MIT) study of Meadows et al. 'The Limits to Growth' from 1972. It was ordered by members of the Club of Rome many of whom were closely related to the OECD Committee on Science and Technology Policy. The science and political strategy of the Club of Rome, which had been founded in 1968, was eventually reinforced by the German Volkswagen foundation, a major research funding organization independent from the car manufacturer, which financed the forecasting study in an effort to assist launching a global debate (Schmelzer, 2012, Bernstein, 2001).

According to Nowotny et al. (2001) a new approach to better integrate science and society has generated good results, generally. Commercialization and increasing pluralism are seen to overcome the longstanding problem of scientific efforts being geared primarily to scientific communication (from ivory tower to ivory tower). Considering the contributions of ecological and environmental policy think tanks one is inclined to endorse the juxtaposition of a new mode of science to the traditional world of more exclusive academia. The old mode has been 'superseded by a new paradigm of knowledge production (Mode II), which was socially distributed, application oriented, trans-disciplinary, and subject to multiple accountabilities' (Nowotny et al., 2003, p. 179). But the closer entanglement with the political and the commercial fields of think tank related activities (Medvetz, 2012) needs to be taken with a grain of salt. If and how and to what end the array of policy expertise is brought to bear to exert influence on the policy process needs to be subject to empirical investigation.

For the purpose of explaining the complications of German energy transition one think tank of the new category discussed here can be considered crucial. In 1994, a cross-party coalition of politicians (CDU, CSU, SPD, Greens) together with forward looking individuals from the corporate sector (like Deutsche Bank chief economist Norbert Walter) founded the association for ecological tax reform (Förderverein ökologische Steuerreform, FÖS), which promoted an increase of taxes on fossil energy sources to be compensated by a reduction of levies on labour. The principal ideas came from the work of the Swiss environmental economist Hans Christoph Binswanger who was among the active scholars of the FÖS. Post-unification Germany was still marked by high levels of unemployment. The politicians and business people active in this group, across the Green, Social Democratic and Christian Democratic party spectrum, therefore aimed at reconciling environmental and economic (employment in particular) policy making. The goal was to create a win-win constellation to unite ecological, economic and social policy goals rather than pitting environmental and economic policy objectives against each other.

The core idea of the new 300-member Think Tank eventually was turned into the political programme of the Social-Democratic-Green coalition government led by Gerhard Schröder. The government introduced a CO_2 tax (exempting coal!) and compensated companies by way of reducing contributions to the social security system (pensions in particular). The programme fell short of the ambitious ecological transformation plan envisioned by the activists of the environmental policy think tanks, but it did manage to keep ecological goals on the agenda and thereby modified the ubiquitous discussion of locational competitiveness in Germany to a certain extent. In addition to the one-time hike of CO_2 taxes (on Diesel and gasoil in particular), the red-green government introduced the renewable energy law in conjunction with nuclear phase out, which featured fixed feed-in tariffs for all the different sources of renewable energy.[4] The renewable energy law since has been reformed numerous times and would become the key concern of the neoliberal and fossil interest groups of the country when the market share of renewables expanded much faster than traditional energy producers had expected. But we will turn to this conflict below and continue the discussion of ecological tax reform first for a moment.

Apart from the effort to reconcile ecology and micro-economic competitiveness, the move towards ecological taxation was also the result of an application of mainstream economics to the environmental challenge in the making since the publication of 'Limits to Growth' (Meadows et al., 1972). The reaction of neoclassical economics to the Club of Rome challenge was defensive at first. The key growth model of the neoclassical synthesis by Robert Solow did not account for finite resource consumption, as consistent with neoclassical tradition (Walker, 2020, p. 19). But Solow and his followers quickly developed an upbeat response: If resources become scarce, the prices will adjust and resource use will be internalized, i.e. become a part of the economic equations calculated in the growth model (Solow, 1973, for an overview of the discussion since see Halkos & Psarianos, 2015). Economists would also be quick to admit that common goods (like clean air and water) need to be brought into the market (models) since free consumption (and pollution leading to additional costs for producers and consumers) are 'external' unless priced, ideally by private contract and the market mechanism, as explained by Ronald Coase in his classical critique of traditional welfare economics (Coase, 1960).

In Germany, the head of the Kiel Institute for World Economy, Herbert Giersch, was a close friend of the MIT 'growth man' Robert Solow and a key protagonist of the move towards 'endogenous growth theory' including the efforts to internalize external cost of common good consumption. Giersch also was a towering figure of German neoliberalism and the president of the Mont Pèlerin Society from 1986 to 1988 (Plehwe & Slobodian, 2019). The efforts Giersch and (some of) his students to integrate the environment into economics from a neoliberal perspective developed momentum. A key collaborator of Giersch in Kiel was Deutsche Bank's Norbert Walter who left Kiel for a new position at Germany's key financial institute where he eventually advanced to become the chief economist. Walter joined a rising star of the Green Party, Sven Giegold, and Social Democrat and Christian Democratic MEPs to set up Förderverein ökologische Steuerreform think tank (FÖS) to advance and apply environmental economics.

The Solow-Giersch-Walter trajectory of embracing environmental economics was both a creative approach of neoliberals to compete in the new world of increasing environmental regulation, and a defense of the neoliberal concern with limited government regulation and market competition at the same time. Green Budget Germany played an important role in this process. 'Each year, it also awards a prize to someone who has been especially helpful in promoting environmental taxation. The award is called the Adam Smith Prize' (Morris & Jungjohann, 2016, p. 167). While the authors are right to emphasize that Adam Smith is here invoked not to fight for lower taxes in the field of environmental policy making, the authors also idealize German neoliberalism (ordoliberalism). They fail to recognize the double-edged sword of neoliberal approaches to environmental economics and the continuing concern of neoliberal environmental economists with growth and competitiveness in opposition to vilified 'command and control' regulation and public ownership. It remains unclear in the field of neoliberal environmental economics, if the major concern is with environmental objectives or with a defensive attitude about free markets in the face of severe (ecological) challenge that may require alternative regulatory and planning approaches. While historical German ordoliberalism has credited the state a strong role to secure a competitive market order in contrast to the second Chicago School (Van Horn, 2009), both American and German neoliberals are united in opposition to Keynesian planning and mixed economy ideas held to compromise market principles (Plehwe, Forthcoming). While supporters of decentralized renewable energy production and a larger number of competitors are keen to invoke ordoliberal ideas in support of competition to legitimize energy transition, modern heirs of ordoliberal ideas have been keener to object to regulating prizes and other command and control (read: public planning) efforts in support of renewable energy production to be discussed below.

The inclusion of key proponents of German neoliberalism in Green Budget Germany in this way is both a success and a liability. The commitment to ecological social market economy carries the neoliberal baggage of the original social market economy ('third way' to prevent socialist/public sector led transformation) and allowed neoliberals to infuse a particular brand of economic reasoning in environmental policy-making, with ambiguous results. One of the applications would soon show in the effort to cut back on the support for renewable energy in Germany, though this can by no means be blamed on the FÖS, which remained an organization committed to advancing environmental policy agendas in spite of the focus on environmental (market) economics. In 2008, the association was officially renamed into Forum Ökologisch-Soziale Marktwirtschaft and became internationally known as the Think Tank 'Green Budget Germany'. We will encounter

Green Budget Germany again in the discussion of the prize of energy transition that has been evoked in order to block further movement towards energy democracy.

4. 'Free market' environmental think tanks: confronting environmental policies and contesting greener neoliberalism

A third group of think tanks has been directed to take up the challenge presented by both ecological and environmental policy think tanks. This third category of explicitly neoliberal think tanks is focused on the nexus between the (unfettered) economic growth paradigm on the one hand, and the calls for more state regulation to protect the natural environment on the other. Pro-active and precautionary environmentalism has been presented (inter alia by Deepak Lal, development economist and Mont Pèlerin president from 2008 to 2010) as a hypertrophic fad of the enemies of the market and as a part of the collectivist conspiracy against the people, a new road to totalitarianism, eco-dictatorship and ecological imperialism (compare Lal, 1998, chapter 6; Lal, 1997). The latter challenge was also taken up in the debate on the 'tragedy of the commons' (Hardin, 1968; for a comprehensive analysis of its impacts, flaws and ideology see Clark, 2010). Students drawing on Hardin in support of regulating the commons frequently overlook his white nationalist ideology and racist political activism.[5] Likewise rarely noticed are Hardin's intentions in favour of full privatization of the commons, the prerequisite for 'Managing the Commons' as the 'promising' alternative to public efforts (Baden & Hardin, 1977). The think tank activities of Hardin's co-editor John Baden, a member of the neoliberal Mont Pèlerin Society like Depaak Lal, are instructive with regard to the employment of new third generation of so-called free market environmental think tanks (Beder, 2001) like the Property and Environment Research Center (PERC) in Bozeman, Montana (founded in 1982), or the Foundation for Research on Economics and the Environment (FREE, founded in 1985, host of the 2004 Mont Pèlerin Society Meeting in Bozeman, Montana) (Plehwe, 2012).

PERC and FREE are part of the neoliberal and corporate countermovement against the agenda developed by the ecological discourse and environmental policy coalitions. In addition to PERC and FREE, the Committee for a Constructive Tomorrow (CFACT) was founded in 1985. This corporate NGO, based in Washington DC, was set up to counter the emerging climate justice movement. The work of neoliberal 'free market' think tanks opposing the sprawling climate justice movement and other challengers has been coordinated to a certain extent by the Atlas Economic Research Foundation based in Virginia, US (Djelic & Mousavi, 2020). Much like the transnational environmental policy elite of the 1970s, these neoliberal circles and corporate sponsors have formed a transnational elite in defense of traditional (fossil) market capitalism, albeit claiming to take environmental concerns into account (Eckersley, 1993). The resulting discourse coalition has not been shy to utilize a wide range of approaches to defend status quo economics ranging from climate change denial to an exclusive emphasis on property rights and market-based instruments in environmental policy making (for see Walker, 2020 for an emphasis on fossil firms in the origins of the neoliberal networks). In Germany much like in the United States and elsewhere, members of the Atlas network include think tanks involved in spreading denial messages and think tanks that do not deny climate change, but aim at preventing ambitious mitigation strategies held to undermine free markets and sound economics.

CFACT has long since its founding in the United States developed a European representation. The European subsidiary of CFACT was founded by the German citizen Holger Thuss in 2004 who was also the founder of the European Institut für Klima und Energie (EIKE, founded in 2007 in

Jena, Germany). Eike has developed into the most active European think tank dedicated to obstructing climate policy, yearly posting hundreds of articles in opposition to climate and energy policies developed in Germany, Europe and in the UN context (Almiron et al., 2020). Eike's work was boosted by the rise of Germany's new right-wing political party Alternative für Deutschland (AFD), which entered both German and European parliaments from 2015. AFD was founded by social conservative and neoliberal economists and is the only party that denies climate change and the need to keep warming below 1.5 degrees. The party thus speaks to the 10–20% of German voters who are not sure about existence and causes of global warming. Stagnating at about 10–12% in election surveys, the new right-wing party is a minor force, but is firmly established in the Germany political spectrum. Beyond denial messages, both AFD and EIKE have a strong track record of objecting to the energy transition and renewables, and instead call for a revival of nuclear energy production. Both CFACT and the German Institut für Unternehmerische Freiheit, EIKE's partner in organizing Germany climate conferences featuring a wide range of denial experts, belong to the Atlas network, though EIKE itself has not been affiliated.[6]

Climate denial is also supported by small fringe groups of the German *mainstream parties*, and at times the AFD/EIKE opposition to energy transition links-up with conservative economic policy forces within the Christian and Social Democratic party. The overlapping constituencies include workers who are adversely affected by the plans to exit coal mining and workers who fear job losses in the German car industry due to the rise of e-mobility, for example (Dörre et al., 2020). But the pro-active forces that overlap with EIKE in the climate and energy political field are not only or even mainly working class related. There are also important corporate partners in the right-wing discourse coalition for a different energy transition primarily in support of 'energy security' and 'locational competitiveness', including the fossil energy producers and large industrial energy customers, as discussed earlier (Michaelowa, 2008, 2013; Stevenson, 2021).

German firms and business associations have been very outspoken in the lobbying controversies around emissions trading. German utilities and industrial firms at first opposed the introduction of European emissions trading in favour of non-binding voluntary commitments to CO_2 reductions. When it became clear that the German government would not be able to block the introduction of the European ETS, German corporations started to reluctantly embrace emissions trading in order to shape the emerging regime to their advantage (Corbach, 2016). In contrast to direct business activism with regard to European emission trading, firms and trade groups have taken a lower profile in the public controversies around renewables, leaving the field to energy economists, think tanks and campaign organizations.[7] Attacks against the highly successful German renewable energy law and the feed-in tariff were stepped-up in the second half of the first decade of the 2000s and have been reinforced after the failure of nuclear revival efforts following the Fukushima accident. Utilities and fossil energy producers for the first time in history were losing money and were starting to shed jobs in Germany and across Europe. Together with industrial customers and energy economists, 'grey' energy trade groups advocated for alternative proposals such as capacity markets and quota models (Haas, 2017, pp. 186–187). These efforts to delay if not derail the Energiewende have taken a number of additional routes, all urging incremental solutions, claiming violations of established rules and emphasizing downsides of the wider energy transformation agenda (compare Lamb et al., 2020 on delay strategies).

A particular European road was taken by the experts of the Center for European Policy in Freiburg (CEP, founded in 2006), an ambitious policy think tank that examines all pieces of European communication and legislation considered relevant to advise constituencies, customers and target audiences on policy issues from a principled neoliberal perspective. CEP did a first analysis of EU plans in support of renewables in 2007 advising against the measures.[8] According to CEP, mandatory goals for a 20% share of renewable energy by 2020 amount to 'downright central planning'.[9] Subsequently, CEP research emphasized contradictions between the stipulations of the German renewable energy law and EU stipulations (Bonn et al., 2014). The study's timing coincided with the European Commission's (EC) case against Germany regarding EEG mechanisms exempting large energy users from cost-sharing for renewables development. These mechanisms were classified as 'state aid' by the Commission. The ECJ ultimately ruled in favour of the German law in 2019 on technical grounds because the redistribution regime from small to large energy customers did not involve direct public spending, but the court case and the argumentation from the CEP experts helped create uncertainty around the renewable energy law and in particular the provision of a feed-in tariff to encourage renewable energy take-up. The prospect of a possible EU determination requiring the repayment of electricity bill savings arising from the feed-in tariff certainly unnerved the industrial customers who had only grudgingly supported the renewable strategy. This role of the CEP effectively as a legal service think tank dedicated to eroding confidence in the energy transformation has so far been neglected in European lobby studies (Günaydin & Plehwe, 2020).

Additional campaigns against the renewable energy law were launched by the Initiative for a New Social Market Economy (INSM, founded in 2000). This campaign organization was set up by the metal employers industry federation in support of neoliberal reforms of the German welfare state and runs on a substantial yearly budget of €7–8 million. INSM attacked the high cost of the alleged 'subsidies' for renewables (INSM, 2012) and advocated switching to a quota model for renewables that was also proposed by the RWI Institute in Essen, a high-end Leibniz-Society research facility (see below). Other government advisory bodies came out in support of switching to the quota model, notably the Monopolies Commission (Monopolkommission, 2013) and the German Council of Economic Experts (Sachverständigenrat, 2014). Last but not least the Think Tank of the German Industry Associations, Institut der deutschen Wirtschaft (IW with offices in Köln and Berlin) published similar positions, which cannot come as a surprise since the IW is the official consultant of the campaign organization INSM. The German feed-in tariff of course had proved to be much more effective than the quota system implemented in UK or Sweden, for example. The feed-in tariff provided a secure investment climate not only for the production of electricity, but also upstream for the producers of machines and equipment. While critics of the feed-in tariff decried the lack of competition due to the regulation of prices (quite common in the traditional fossil energy market), the regime can be considered to acknowledge ordoliberal ideas pertaining to state efforts to ensure a competitive market: it has been successful in creating new forms of local competition for renewable energy in what had been a highly centralized oligopolistic fossil fuel energy market. Secure returns, put in place through the feed-in tariff, helped support the emerging renewable energy market, creating a more equal playing field for a very diverse group of competitors. Despite this, spurious (and self-interested) claims that price fixing was necessarily anti-competitive, made by the dominant incumbent firms and their allies in academic and think tank circles, were effective in shaking confidence, and slowing down the growth of renewables.

The role played by the academic research institute RWI is particularly worrying. With the establishment of its 'Environment and Energy' department, under the leadership of Manuel Frondel, the RWI allowed this energy economist to launch a campaign against the renewable energy law and the feed-in tariff. Between 2006 and 2011, the institute published at least seven studies, which discussed the alleged inefficiency of renewables and state support. In 2010, Frondel presented an RWI study calculating the cost of supporting renewables at 64 billion Euros, for example, whilst failing to account for the benefit of falling electricity prices with the rising share of renewables, and other aspects.[10]

Arguably the biggest flaw in the RWI study was the failure to mention the sponsor of the study, the U.S. Institute for Energy Research (IER), which is a climate change denial think tank funded by oil and coal corporations including the Koch family. The link was uncovered by German TV format Monitor in 2010 (Kemfert 2013, p. 83). This scandal did not have a negative impact on Frondel's academic career as an economist and energy expert, however. He continues to work jointly with the industry campaign organization INSM, lately weighing-in for emission trading and against CO_2 taxes, for example.[11] While organizations like RWI and INSM generally keep their distance from the local climate denial think tank in Germany, the EIKE, the transnational obstruction strategy has clearly been effective in recruiting reputable neoliberal academics in institutes like RWI where it is important to defend the status quo against the Energiewende and wider pressures for Energy Democracy.

Emphasizing the allegedly high cost of the feed-in tariff and the resulting burden for the average German citizen (6.5 cents per kwh in 2021) was another cornerstone in the campaign against the feed-in tariff. In 2013, preceding the 2014 EEG reform that got dismantled the feed-in tariff, Germany's economics minister Peter Altmaier jumped on the 'energy poverty' bandwagon, publishing fantasy numbers of the future cost of continuing the feed-in tariff. With this the neo-liberal campaign against the feed-in tariff entered the field of propaganda: Green Budget Germany was one of the organizations that revealed that in fact there was no study supporting the figures that were cited by the campaign organizations and the minister (Morris & Jungjohann, 2016, p. 366). Both the CEP in Freiburg and the corporate campaign organization INSM are part of the global Atlas network. While not dedicating efforts to global warming denial, the evidence of opposition to ambitious mitigation efforts in the name of competitiveness and efficient policy making is overwhelming.

Ending the feed-in tariff in 2014 led to the decline in the growth rate of solar and wind energy production in Germany. The simultaneous introduction of new rules for wind turbines, including requirements for them to be distanced from settlements, added additional regulatory constraints on the expansion of renewable energy. In 2019 the German Coal Commission decided that all German coal-fired plants should close by 2038 (which itself unlikely to be compatible with the Paris agreement). Yet the following year a brand-new 1.1 GW coal fired power station went online in Datteln, maintaining coal emissions for decades.

While it has been impossible to stop the energy transition, industry interests and neoliberal allies have succeeded in delaying the expansion of renewables and the phasing out of fossil fuels. Shaping the process in this way, effectively to lengthen the transition period, is critically important from the perspective of the incumbent firms. Blocking the rise of decentralized renewable production, onshore wind and solar in particular, restrains the threat of sources of competition, while capital-intensive renewables energy production especially in the form of offshore wind parks, can be developed by the incumbent energy sector firms. To close the circle, opposition to onshore wind energy has been mobilized by EIKE and AFD constituencies in conjunction with conservative environmental groups and front groups funded by commercial interests (Redelfs, 2021; Kwasniewski, 2021). The preferred neoliberal energy 'transition' stalls distributed renewable energy, prolongs fossil fuels where possible, and (re)captures energy markets for the utility-scale renewables run by the existing energy corporates.

A key aspect of the strategy is the turn towards neoliberal environmental economics and market instruments. It has already been mentioned briefly that energy economist Frondel of RWI supported the INSM campaign in favour of emission trading and in opposition to CO_2 taxes, which are both economic instruments. CO_2 taxes of course have a more direct impact by clearly adding a government-calculated price tag to CO_2 emissions, whereas emission trading produces a much more uncertain outcome in terms of pricing and emissions reduction. Of course, the government has a strong interest in policy that is most likely to deliver emissions reduction, not least as reduction targets are set by the state. This ultimately explains the government's preference for a tax-based carbon pricing system price rather than the trading mechanism. In contrast, emissions trading is strongly preferred by industrial interests as a 'market' mechanism that will minimize carbon prices. Within the European Emissions Trading System energy corporates were able keep the carbon price low (and ineffective) by using the free certificates given to incumbent emitters, and through many other means of negating the market instrument (including international offsets, joint implementation etc., compare Gilbertson & Reyes, 2010, 55f.).

Most important for the discussion of the renewable energy law, economists have argued that there is no longer a need for a dedicated regime in support of energy conversion since emission trading allegedly takes care of all the necessary transformation processes. The renewable energy law is held to be superfluous and even counterproductive since it interferes with the 'market', for instance by effectively lowering the price for CO_2 certificates (Frondel & Schmidt, 2006). In making this argument, our RWI authors omitted the fact that CO_2 reductions achieved through the EEG regime were reduced from the overall CO_2 budget to take care of such perverse effects and to secure the effectiveness of the regime (Kemfert & Diekmann, 2009).

5. Conclusions: status quo ante and preparing for the next stage of energy conversion

This chapter focused on three different categories of think tanks and related discourse coalitions to examine and explain the present conundrum of German energy transition and climate politics: Energiewende introduced by the Öko-Institut, environmental modernization promoted by Green Budget Germany and neoliberal 'environmental' economics exemplified by CEP in Freiburg and the employer campaign organizations INSM, for example. Germany is rich in all three segments of think tanks, ecological, environmental modernization and neoliberal economic approaches to environmental questions, which all link competing varieties of academic expertise, communication and policy capacities, and interest and advocacy groups. Although Germany was early to introduce new ideas and policy proposals in support of the transition to renewable energy and energy democracy through think tanks and citizen movements, the rise in the share of renewables has been slowed in the 2010s and the trajectory of Energiewende has been stalled. As argued here, the project has been derailed by a combination of moderate and reactionary influences emphasizing economic competitiveness concerns. While climate denial efforts by organizations like EIKE play a role in the far-right spectrum of German politics mostly, the mobilization of neoliberal perspectives to curtail regulatory approaches like the feed-in-tariff proved successful to stall the expansion of renewables in the second half of the 2010s. The article illustrates the capacity of the neoliberal brand of 'environmental' economics and think tanks that aim at aligning neoliberal concerns with competitiveness and environmental policy demands. These efforts play out to subordinate ecological goals to economic goals, competition and growth. While ecological modernization strategies advanced by organizations like Green Budget Germany place a much stronger emphasis on environmental policy goals and can still be considered to be in conversation with more radical ecological reform movements, neoliberal environmental economics has nevertheless become a dominant discourse utilized to delay the exit from fossil fuels and to block the movement towards decentralized energy production. It appears to be critical for the future of environmental economics to more clearly distinguish between genuine environmental economics and economics only nominally concerned with environmental policy goals. The claims of economists to have the one and only 'silver bullet' approach to deal with environmental concerns in any case needs to be soundly rejected in support of a spirit of trans- and interdisciplinary collaboration concerning ecological needs with economic concerns being just one of many rather than claiming the hierarchical top. Future research on the ecological modernization projects like the European Green Deal needs to address the limits of European competition and state aid policy, for example, in order to enable a wider range of policy options to enable ambitious mitigation and energy democracy.

Notes

1. The Social-Democrat Green Coalition terminated the end of nuclear energy in 2000 in conjunction with the commitment to rapidly expand renewable energy, but the Christian-Democrat-Liberal coalition government reversed the decision in 2010. The Fukushima meltdown eventually led chancellor Merkel to reconsider the extension of nuclear energy production returning to the original phaseout schedule (Radkau & Hahn, 2013).
2. Konrad's father Helmuth James Graf von Moltke was a member of the Kreisauer Kreis behind the failed coup. His mother Freya escaped to South Africa with her children and later moved to the United States. Konrad's history and work is recalled by the German Ecologic Institute, which was inspired by the work of von Moltke (see https://www.ecologic.eu/1401).
3. As confirmed in interviews with founder of Ecologic, R. Andreas Kraemer, and with Udo Ernst Simonis, former director of environmental policy research unit at the Berlin Social Science Center, conducted by the author
4. An overview of the record of the two red-green coalition governments in the environmental policy field is provided here: https://www.upi-institut.de/bilanz.htm (last accessed 14 May 2021).
5. An excellent review of the two sides of Hardin's writing has been provided by the Southern Poverty Law Center: https://www.splcenter.org/fighting-hate/extremist-files/individual/garrett-hardin, last accessed 26 October 2021.
6. Atlas has removed its detailed directory from the website (https://www.atlasnetwork.org/partners), but information can be retrieved up to Spring 2017 via achrive.org.
7. All the firms and business associations were involved in multi-level lobbying strategies within the institutions. Compare Ydersbond (2012) for details of the 'economic' and the 'renewable' coalition and numerous internal divisions.
8. It is unclear if and to which extent funding for CEP has been secured from fossil interest groups. A major economic link of CEP is at evidence with the co-chair of the Council, Christoph Ehlers, who held positions in banking and runs a legal consulting organization to support startups. The website does not include information on funding that can be conservatively estimated in the range of €2–3 million per year for the four locations in Freiburg, Berlin, Paris and Rome (based on staff salary).
9. https://www.cep.eu/fileadmin/user_upload/cep.eu/Analysen_KOM/KOM_2006_848_Fahrplan_erneuerbare_Energien/cepAnalyse_KOM_2006_848_Fahrplan_fuer_erneuerbare_Energien.pdf.
10. https://www.photovoltaik.eu/article-444463-30021/die-luege-vom-teuren-oekostrom-.html.
11. https://www.insm.de/insm/presse/insm-veranstaltungen/2019/fruehstuecksdialog-klima-und-umweltpolitik.

Disclosure statement

No potential conflict of interest was reported by the author(s).

ORCID

Dieter Plehwe http://orcid.org/0000-0002-8563-0946

References

Almiron, N., Boykoff, M., Narberhaus, M., & Heras, F. (2020). Dominant counter-frames in influential climate contrarian European think tanks. *Climatic Change, 162*(4), 2003–2020. https://doi.org/10.1007/s10584-020-02820-4

Baden, J., & Hardin, G. (Eds.). (1977). *Managing the Commons*. W. H. Freeman and Company.

Becker, S., Beveridge, R., & Röhring, A. (2016). Energy transitions and institutional change: Between structure and agency. In L. Gailing, & T. Moss. (Eds.), *Conceptualizing Germany's energy transition: Institutions, materiality, power, space* (pp. 21–41). Springer. https://eui.idm.oclc.org/login?url=http://doi.org/10.1057/978-1-137-50593-4

Beder, S. (2001). Research note – Neoliberal think tanks and free market environmentalism. *Environmental Politics, 10*(2), 128–133. https://doi.org/10.1080/714000530

Bernstein, S. (2001). *The compromise of liberal environmentalism*. Columbia University Press.

Bonds, E. (2011). The knowledge-shaping process: Elite mobilization and environmental policy. *Critical Sociology, 37*(4), 429–446. https://doi.org/10.1177/0896920510379440

Bonn, M., Heitmann, N., Reichert, G., & Voßwinkel, J. S. (2014). *Entwurf der Leitlinien der Europäischen Kommission für staatliche Umwelt- und Energiebeihilfen 2014–2020*. Centrum für Europäische Politik.

Böcher, M., & Töller, A. E. (2011). *Umweltpolitik in Deutschland. Eine politikfeldanalytisch orientierte Einführung* (Kurs Nr. 03900 der FernUniversität Hagen).

Brulle, R., & Aronczyk, M. (2019). Environmental countermovements: Organised opposition to climate change action in the United States. In A. Kalfagianni, D. Fuchs, & A. Hayden (Eds.), *Handbook of global sustainability governance* (pp. 218–230). Routledge.

Clark, J. (2010). The tragedy of common sense part one: The power of myth. *Capitalism Nature Socialism, 21* (3), 35–54. https://doi.org/10.1080/10455752.2010.507501

Coase, R. H. (1960). The problem of social cost. *The Journal of Law and Economics, 3*, 1–44. https://doi.org/10.1086/466560

Corbach, M. (2016). *Energiepolitischer Lobbyismus in Deutschland. Eine Fallanalyse zur Einführung des Emissionshandels*. Freie Universität Berlin (Dissertation).

Dörre K., Holzschuh M., Köster J., & Sittel J. (Eds.). *Abschied von Kohle und Auto? Sozial-ökologische Transformationskonflikte um Energie und Mobilität*. Campus.

Eckersley, R. (1993). Free market environmentalism: Friend or foe? *Environmental Politics, 2*(1), 1–19. https://doi.org/10.1080/09644019308414061

Engfeldt, L.-G. (2009). *From Stockholm to Johannesburg and beyond*. Government Offices of Sweden, Ministry of Foreign Affairs.

Fischer, F. (1993). Policy discourse and the politics of Washington think tanks. In F. Fischer & J. Forrester (Eds.), *The argumentative turn in policy analysis and planning* (pp. 21–42). Duke University Press.

Frondel, M., & Schmidt, C. M. (2006). *Emissionshandel und Erneuerbare-Energien-Gesetz: Eine notwendige Koexistenz? RWI Positionen*. RWI Essen. http://www.rwi-essen.de/media/content/pages/publikationen/rwi-positionen/Pos_010_Emissionshandel.pdf

Geels, F. W. (2014). Regime resistance against low-carbon transitions: Introducing politics and power into the multi-level perspective. *Theory, Culture & Society, 31*(5), 21–40. https://doi.org/10.1177/0263276414531627

Gilbertson, T., Reyes, O., & Peinelt, E. (2010). *Globaler Emissionshandel: Wie Luftverschmutzer belohnt werden - Analyse, Kritik, Perspektiven*. Brandes & Apsel.

Günaydin, K., & Plehwe, D. (2020). *The neoliberal legal turn. The Center for European Politics (CEP) and its new type of legal service provision*. http://thinktanknetworkresearch.net/blog_ttni_en/the-neoliberal-legal-turn/

Haas, T. (2017). *Die politische Ökonomie der Energiewende: Deutschland und Spanien im Kontext multipler Krisendynamiken in Europa*. Springer VS.

Hajer, M. A. (1993). Discourse coalitions and the institutionalization of practice: The case of acid rain in Britain. In F. Fischer & J. Forrester (Eds.), *The argumentative turn in policy analysis and planning* (pp. 43–76). Duke University Press.

Halkos, G., & Psarianos, I. (2015). *The effect of including the environment in the neoclassical growth model* (MPRA paper no. 68050). https://mpra.ub.uni-muenchen.de/68050/

Hardin, G. (1968). The tragedy of the commons. *Science,* New Series *162*(3859), 1243–1248.

Hirschl, B., & Vogelpohl, T. (2019). Energiepolitik in Deutschland und Europa. In J. Radtke & W. Canzler (Eds.), *Energiewende. Eine sozialwissenschaftliche Einführung* (pp. 69–95). Springer.

INSM. (2012). *Aspekt 2: Wirtschaftlichkeit.* Updated November 5, 2012. Retrieved January 23, 2020, from https://www.insm.de/insm/themen/soziale-marktwirtschaft/energiewende-radar/aspekt2-wirtschaftlichkeit

Kemfert, C. (2013). *Kampf um Strom. Mythen, Macht und Monopole.* Murmann.

Kemfert, C., & Diekmann, J. (2009). Förderung erneuerbarer Energien und Emissionshandel: wir brauchen beides. *DIW, 27*(11), 169–174. https://www.diw.de/documents/publikationen/73/diw_01.c.96064.de/09-11-3.pdf

Krause, F., Bossel, H., & Müller-Reißmann, K.-F. (1981). *Energiewende: Wachstum und Wohlstand ohne Erdöl und Uran.* S. Fischer.

Kwasniewski, N. (2021) *Netzwerk mit Unterstützung der Industrie: Die Anti-Windkraft-Bewegung.* SPIEGEL Wirtschaft. https://www.spiegel.de/wirtschaft/windenergie-so-verhindert-die-anti-windkraft-bewegung-neue-anlagen-a-46d88419-3b1d-427d-b6c0-cf696fec283c.

Lal, D. (1997). Ecological imperialism. In J. Morris (Ed.), *Climate change – Challenging the conventional wisdom.* Institute of Economic Affairs (EA Environment Unit). http://www.policynetwork.net/uploaded/pdf/climate_change_book.pdf

Lal, D. (1998). *Unintended consequences.* MIT Press.

Lamb, W. F., Mattioli, G., Levi, S., Timmons Roberts, J., Capstick, S., Creutzig, F., Minx, J. C., Müller-Hansen, F., Culhane, T., & Steinberger, J. K. (2020). Discourses of climate delay. *Global Sustainability, 3,* e17, 1–5. https://doi.org/10.1017/sus.2020.13

Lovins, A. B. (1977). Energy strategy: The road not taken? In *Foreign affairs. Soft energy paths. Towards a durable peace.* Harper (Original work published 1976).

McCormick, J. (1989). *Reclaiming paradise: The global environmental movement.* Indiana University Press.

Meadows, D. H., Meadows, D. L., Randers, J. & Behrens III, W.W. (1972). *The Limits to Growth.* Universe Books.

Medvetz, T. (2012). *Think-tanks in America.* University of Chicago Press.

Meyer, J.-H. (2011). *Appropriating the environment. How the European institutions received the novel idea of the environment and made it their own* (KFG working paper 31). Free University Berlin.

Michaelowa, A. (2008). German climate policy between global leadership and muddling through. In H. Compston (Ed.), *Turning down the heat* (pp. 144–163). Palgrave.

Michaelowa, A. (2013). The politics of climate change in Germany: Ambition versus lobby power. *WIREs Climate Change, 4*(4), 315–320. https://doi.org/10.1002/wcc.224

Monopoly Commission 'Sektorgutachten Energie'. (2013). *Wettbewerb in Zeiten der Energiewende (ehem. SG 65).* German. https://www.monopolkommission.de/de/gutachten/sektorgutachten-energie/114-sektorgutachten-energie-4.html

Morris, C., & Jungjohann, A. (2016). *Energy democracy: Germany's Energiewende to renewables.* Palgrave Macmillan.

Morton, T., & Müller, K. (2016). Lusatia and the coal conundrum: The lived experience of the German Energiewende. *Energy Policy, 99,* 277–287. https://doi.org/10.1016/j.enpol.2016.05.024

Moss, T., & Gailing, L. (2016). Introduction. In L. Gailing, & T. Moss. (Eds.), *Conceptualizing Germany's energy transition: Institutions, materiality, power, space.* Springer. https://eui.idm.oclc.org/login?url=http://doi.org/10.1057/978-1-137-50593-4

Nowotny, H., Scott, P., & Gibbons, M. (2001). *Re-thinking science: Knowledge and the public in an age of uncertainty.* Polity Press.

Nowotny, H., Scott, P., & Gibbons, M. (2003). Introduction. 'Mode 2' Revisited: The New Production of Knowledge. *Minverva, 41*(3), 179–194.

Plehwe, D. (2012). 'Umwelt-think tanks' – Denker oder Lenker. In H. Leitschuh, G. Michelsen, U. E. Simonis, J. Sommer, & E. U. von Weizsäcker (Eds.), *Wende überall?* (pp. 93–100). Jahrbuch Ökologie 2013.

Plehwe, D. (2015). The politics of policy think-tanks: Organizing expertise, legitimacy and counter-expertise in policy networks. In F. Fischer, D. Torgerson, A. Durnová, & M. Orsini (Eds.), *Handbook of critical policy studies* (pp. 358–379). Edgar Elgar.

Plehwe, D. (Forthcoming). The economic state and the ordoliberal critique of Keynesianism. Anti-economic state or just a different type of economic state? In T. Biebricher, P. Nedergaard, & W. Bonefeld (Eds.), *The Oxford handbook of ordoliberalism*. Oxford University Press.

Plehwe, D., & Slobodian, Q. (2019). Landscapes of unrest: Herbert Giersch and the origins of neoliberal economic geography. *Modern Intellectual History*, *16*(1), 185–215. https://doi.org/10.1017/S1479244317000324

Radkau, J., & Hahn, L. (2013). *Aufstieg und Fall der deutschen Atomwirtschaft*. oekom.

Redelfs, M. (2021). *Die Gegner der Energiewende*. Greenpeace Recherche. www.greenpeace.de/files/publications/gegner_der_windkraft.pdf

Renewables. (2021). *Global status report*. Retrieved October 16, 2021, from https://www.ren21.net/wp-content/uploads/2019/05/GSR2021_Full_Report.pdf

Roose, J. (2002). *Made by Öko-Institut*. Öko-Institut.

Rüb, F. W. (Ed.) (2014). *Rapide Politikwechsel in der Bundesrepublik*. Nomos.

Rucht, D. (1980). *Von Wyhl nach Gorleben. Bürger gegen Atomprogramm und nukleare Entsorgung*. C.H. Beck.

Sachverständigenrat. (2014). Chapter 10 in Energiepolitik: Warten auf die dringendnotwendigen Weichenstellungen. In *Jahresgutachten 2013/2014*.

Salles-Djelic, M. L., & Mousavi, R. (2020). How the neoliberal think tank went global: The atlas network, 1981 to the present. In D. Plehwe, Q. Slobodian & P. Mirowski (Eds.), *Nine lives of neoliberalism* (pp. 257–282). Verso.

Saloma, J. S., III. (1984). *Ominous politics. The new conservative labyrinth*. Hill and Wang.

Scheydt, S. (2007). Wandel durch Annäherung. *Brand eins*, *9*(12), 38–46.

Schmelzer, M. (2012). The crisis before the crisis: The 'problems of modern society' and the OECD, 1968–74. *European Review of History: Revue europeenne d'histoire*, *19*(6), 999–1020. https://doi.org/10.1080/13507486.2012.739148

Schmitter, P. C., & Streeck, W. (1999). *The organization of business interests: Studying the associative action of business in advanced industrial societies* (MPIfG discussion paper 99/1). Max Planck Institute For the Study Of Societies.

Smith, J. A. (1991). *The Idea Brokers: Think Tanks and the Rise of the New Policy Elite*. Free Press.

Solow, R. M. (1973). Is the end of the world at hand? *Challenge*, *16*(1), 39–50. https://doi.org/10.1080/05775132.1973.11469961

Spangenberg, J. (2012). NGOs between influence and participation overkill. In O. Renn, A. Reichel, & J. Bauer (Eds.), *Civil society for sustainability – A guidebook for connecting science and society* (pp. 5–17). EHV.

Stefes, C. H. (2014). Energiewende: Critical junctures and path dependencies since 1990. In F. W. Rüb (Ed.), *Rapide Politikwechsel in der Bundesrepublik* (pp. 47–70). Nomos.

Stevenson, H. (2021). Reforming global climate governance in an age of bullshit. *Globalizations*, *18*(1), 86–102. https://doi.org/10.1080/14747731.2020.1774315

Stone, D. (2007). Garbage cans, recycling bins or think-tanks? Three myths about policy institutes. *Public Administration*, *85*(2), 259–278. https://doi.org/10.1111/j.1467-9299.2007.00649.x

Strong, M. (2000). *Where on Earth are we going?* Texere.

Van Horn, R. (2009). Reinventing monopoly and the role of corporations: The roots of Chicago law and economics. In P. Mirowski & D. Plehwe (Eds.), *The road from Mont Pèlerin* (pp. 204–237). Harvard University Press.

Walker, J. (2020). *More heat than life: The tangled roots of ecology, energy, and economics*. Palgrave MacMillan.

Weidner, H. (1996). *Umweltkooperation und alternative Konfliktregelungsverfahren in Deutschland. Zur Entstehung eines neuen Politiknetzwerkes*. Wissenschaftszentrum Berlin für Sozialforschung.

Ydersbond, I. M. (2012). *Multi-level lobbying in the EU: The case of the renewables directive and the German energy industry*. Fridtjof Nansen Institute.

Energy transition and dialectics: tracing discursive resistance to coal through discourse coalition in India

Rajshri Shukla and Pradip Swarnakar

ABSTRACT

The hegemonic narrative linking coal to national sovereignty and energy security promoted by state and private actors, translates into policies promoting coal expansionism. The study is based on the proposition that the counter-discourse offered by non-state actors, mainly environmental civil society organization, represents a potential challenge against such narrative. The dialectical discursive dynamics around energy transition are explored within India's print media by employing the method of Discourse Network Analysis to trace the dominant storylines and discourse coalitions. The findings reveal two opposed storylines, one pro-coal the other anti-coal, underpinning the two ideologically cohesive discourse coalitions dominated by state and non-state actors, respectively. It uncovers the ideological fault lines between the two sets of actors, with the possibilities for synthesis and policy co-learning. The paper concludes by deriving implications regarding post pandemic challenges and possibilities emerging for both the sets of actors.

Introduction

About two-thirds of the emissions causing climate change come from the energy sector, and therefore, it has to be central to any mitigative action (Lee & Birol, 2020). As a result, energy policy is becoming increasingly entwined with climate change discourse, as energy transition[1] away from fossil fuels presents itself as the most obvious solution to the ongoing climate crises. The process faces challenges in the form of deep lock-in of fossil fuel especially coal in global economic order (Urry, 2014). As a result, actors with vested socio-economic interest devise and resort to myriad ways to sustain coal usage. One of these ways operates at the level of discourse making where discursive strategies are deployed by the incumbent actors to justify the use of coal (Buschmann & Oels, 2019; Černý & Ocelík, 2020; Schneider et al., 2016).

Such discursive practices play an influential role in sustaining and promoting coal in coal dependent countries like India, the focus case of the current study, where it is central to energy needs for multiple reasons ranging from natural abundance, ensuring energy security and politico-economic drivers (Dubash et al. 2018; Lahiri-Dutt, 2014; Talukdar, 2016). There is a real dilemma faced by India in terms of climate induced phasing out of coal when this cheap, abundant source can provide an easy solution to the pressing energy poverty issues in the country. Consequently, the country has seen a two-sided tendency of simultaneous coal and renewable expansion (Roy & Schaffartzik, 2021). Such tendency results in a dialectical conflict at discursive level where various actors like state authorities, businesses, civil society and research organizations frame their pro and anti-coal narratives (Swarnakar et al., 2021).

The aim of this study is to delve deeper into this discursive dialectical process to answer two questions – What are the main themes emerging out of this dialectical discourse? And, which actors engage in discursive conflict over coal phase-out entailing energy transition and what is the potential of the actors resisting coal? The study particularly relies on media discourse and extends the theoretical concepts of storylines and discourse coalitions (Hajer, 1993) and 'mere possibility' (Bloch, 1986) to grasp better the promoted themes, interaction of various contending actors in the media and the potential of the anti-coal actors. For examining such interaction, we take a network approach to analyzing media discourse by employing the method of Discourse Network Analysis (DNA) (Leifeld, 2017).

The contribution of our study lies in three aspects. First, we empirically delineate the dominant storylines and discern pro-coal and anti-coal coalitions within the overall energy transition discourse in a coal dependent country which has a major role in the global fight against climate change. Second, taking a cue from Gunderson's (2020) project of finding merely possible solutions within the dialectics of climate change, we make the proposition that the discursive resistance offered by anti-coal discourse coalition offers a 'mere possibility' (Bloch, 1986) of overcoming fossil fuel dependency, and through our analysis provide evidence for this. Third, it spells out the possible policy trajectories that may emerge in terms of energy transition and demonstrates the potential legitimacy that they may have in terms of support from various actors.

The next section contextualizes our country-case by elaborating India's coal dependence and its impacts on energy transition. The following section discusses the theoretical underpinnings of our research and their analysis in the form of dialectics, discourse, and discourse coalitions. The methods section of the paper elucidates how we have operationalized the various aspects of discourse coalitions. Finally, we demonstrate our results and identify their implications.

The Indian energy transition: the diverging roads

The post-independence nation-building narrative in India hailed coal as the backbone of economic development and closely connected it with the sovereignty of the state (Lahiri-Dutt, 2014). In 2020, it continues to be the central energy source with 40% of primary energy demand being met by coal. It constitutes the single largest element in the energy mix and fuels the energy access expansion in the country (International Energy Agency, 2021). The coal rich regions of the country comprising of eastern states like Jharkhand, Odisha, Chhattisgarh and West Bengal have predominantly a coal-based economy with more than 13 million people dependent on coal and allied sectors across the country (Dsouza & Singhal, 2021).

The energy sector in general and coal in particular are largely state owned. Montrone et al. (2021) provide a useful insight into the actors based political economy of coal in the country. The actors like the Prime Minister's Office (PMO), Ministry of Power (MoP), Ministry of Coal (MoC), Ministry of Petroleum and Natural Gas (MoPNG) play a major role in the overall governance from extraction and generation to distribution of power. State-owned Coal India Limited (CIL) dominates the extraction and production of coal. With mounting urgency around environmental and climate concerns, the Ministry of New and Renewable Energy (MNRE) and Ministry of Environment, Forest and Climate Change (MoEFCC) have gained prominence in the energy policy making are respectively responsible for deployment of non-conventional energy sources like solar, wind, biogas and giving environmental clearances for energy and mining projects.

Despite state dominance, the multiple attempts at liberalizing the coal sector since 1992 have resulted in an enhanced role of the private players. In 2022, the privately owned coal based installed capacity stands close to 36% of the total coal based installed capacity (CEA, 2022). The majority of this share is concentrated within a few large companies like Adani, Tata, and Reliance (Montrone et al., 2021) and produces the 'neoliberal coal' economy (Lahiri-Dutt, 2016). This has extended the 'coal geography' to the coastal regions of the country with grave socio-environmental consequences (Oskarsson et al., 2021).

The vested interests in the coal economy, therefore, range from those of political and economic elites to the vulnerable coal communities and workforce across coal regions. While the former exploit coal reserves for revenue and profits and advance the discourse of national future energy security to justify coal (Swarnakar et al., 2021), the latter rely on coal for subsistence albeit precarious, with devastating outcomes for coal communities in terms of health and safety hazards and adverse socio-economic conditions (Ahmad & Lahiri-Dutt, 2006; Lahiri-Dutt, 2014).

With such dependence ranging across multiple stakeholders, the trajectory of climate change induced energy transition away from coal and other fossils is made uncertain. The country made ambitious renewable commitments at the United Nations Framework Convention for Climate Change (UNFCCC) Conference of Parties (COP). The target of 175 GW of renewable energy till 2022 made in COP 21, Paris in 2015 (MoEFCC, 2015) was enhanced to 500 GW by 2030 in COP 26, Glasgow in 2021 (Ministry of External Affairs, 2021).

Despite such ambitious commitments a clear roadmap for a dedicated policy to govern a just transition away from fossil fuels remains absent (Jairaj & Kumar, 2019). Even with renewables becoming increasingly economically competitive and feasible, government policies have retained a focus on coal expansion (Dubash et al. 2018; Goodman, 2016; Roy & Schaffartzik, 2021).

As part of the pandemic recovery package released in 2020, the government announced the auction of concessions for fifty new coal mines each with a life of about 50 years (Behl, 2020). This move will create a future carbon lock-in in a country which is the third largest emitter globally resulting in adverse consequences for the global fight against climate change, especially for the most 'vulnerable populations' (Dubash, 2019, p. 3). A counter-discourse to unsustainable coal dependence has come from civil society organizations (CSO) like The Energy and Resources Institute (TERI). Such CSOs have been influential in articulating a 'co-benefits' approach to climate policy, where the benefits of moving to renewable energy are emphasized. This has encouraged more climate-friendly and renewable oriented policies in the past (Aamodt & Stensdal, 2017; Mayrhofer & Gupta, 2016).

Given this context, it becomes urgent to more systematically explore for possibilities of various avenues from where the resistance to coal expansionism can emerge. We look for such possibilities of resistance within the dialectical interactions between the discourse coalitions emerging out of the Indian energy transition discourse.

Energy, dialectics[2] and discourse coalition

To bring energy into the ambit of sociological enquiry it has to be conceptualized as a 'social relation embedded in dense networks of power and socioecological change': this can allow a 'perspective of dialectical complexity', of conflict as well as co-production between nature and society (Huber, 2009, p. 106). Drawing also on Moore's discussion of nature/society dualisms (2015) we can interpret energy as being increasingly bound into the logic of a 'climate dialectic' (Goodman, 2016, p. 184). Energy policy is increasingly drawn into this dialectical relationship through 'a series of tensions and contradictions [that] open up under climate change, focused especially on … how to transform energy supply systems'. (Goodman, 2016, p. 184). While unfolding, the dialectical conflicts that emerge within the energy system under climate change give rise to divisions among social groups, making energy the 'great divider' (Urry, 2014, p. 9).

In a more recent reflection, Gunderson (2020), suggests the theoretical-analytical category of 'negative dialectics' to better analyse emissions-reduction efforts proposed to address climate change. The approach can help in bringing-out the 'mere possibilities' (Bloch, 1986, p. 231) of achieving better social conditions, inherent in the existing governing systems of climate change mitigation. In its conservative use, the method of negative dialectics seeks to uncover the conditions that prevent human emancipation and discount the possibility of revolutionary change (Gunderson, 2020, p. 606). However, Bloch's conception of 'mere possibility', as Gunderson argues, 'clears space for affirmative or positive speculation that negative dialectics avoids' (Gunderson, 2020, p. 613). Here, it is pertinent to visit Bloch's distinction between the internal and external dimension of the category of 'Possible'. The internal dimension or *capacity* is the subjective human ability to change while the external dimension or the *potentiality* are the external conditions referring to 'social, technical, and environmental factors that are partially ripe or ripe for change' (Gunderson, 2020, p. 611). The scope for the merely possible 'remains even if one of the two conditions, the internal or the external, should almost be fulfilled (Bloch, 1986, p. 231). In other words, the term 'mere possible' is used for conditions that may either have capacity but no potentiality or have favourable external potentiality while lacking a capacity in terms of weak-willed human subjects.

The study seeks to trace such possibilities in the conflicts related to energy that unfold, as a dialectical process through mapping the discourse around energy transition. In terms of policy-making in general and around climate change and energy in particular, 'discourses define the range of policy outcomes and operate as resources which empower certain actors and exclude others', while also serving as 'sites of resistance, fomenting the emergence of counter-discourses' (Litfin, 1995, p. 253). Media discourse on climate and energy governance, with various claims and claim-makers (Boykoff, 2011) offer such 'publicly visible sites for conflict' (Stoddart & Tindall, 2015, p. 402). A significant body of literature explores the media representation of energy transition to demonstrate that the socio-political and structural conditions influence the ways in which the mass media frames renewable energies (Djerf-Pierre et al., 2016; Rochyadi-Reetz et al., 2019). Other literature focuses on the coal phase-out to highlight its contested nature, with various actors engaging in a framing struggle within the media (Isoaho & Markard, 2020). There has also been research into the discursive practices of identity reconstruction and justice within the process of energy transitions (Ganowski & Rowlands, 2020; Walker et al., 2019) and explorations into the formation and changes in the discourse coalitions around energy transition (Hess, 2019).

The framework of discourse coalitions provides middle-range concepts (Hajer, 1995, p. 61) to make theoretical sense of real-world conflicts. It defines discourse as a shared system of ideas and concepts that gives meaning to phenomena along with selectively highlighting some aspects of the problem by framing it in a certain way (Hajer, 1993, p. 45). Discourse coalitions emerge through this shared process of meaning-making, defined as 'ensemble of a set of story lines, the actors that utter these storylines, and the practices that conform to these storylines, all organized around a discourse' (Hajer, 1993, p. 47).

The actors in these coalitions are connected through shared storylines[3] instead of actual strategizing (Fischer, 2003), thereby diverging from the common sense understanding of the term. However, previous studies have shown that even though the actor and storyline composition of the coalitions may change over time (Hess, 2019) or they may exhibit plurality in terms of storylines and interests (Jessup, 2010), the discursive constructions employed by discourse coalitions are crucial in the sense that they enable or constrain policy options regarding climate change (Bulkeley, 2000). Riedy (2020), explores this discursive power of discourse coalitions against the unsustainable neoliberal narratives by identifying the common ground that can provide converging storylines for actors to come together to push for sustainable systems.

In the Indian context, a successful and just transition to sustainable energy systems is largely hindered by the dominant narrative around the importance of coal. To demonstrate this, we identify the key storylines that dominate the media discourse on energy transition away from coal, and discern discourse coalitions based on conflicting storylines on coal in the Indian energy transition and through this look for 'seeds of transformation' (Riedy, 2020, p. 108) in the discursive resistance against coal.

Materials and method

The current study employs the method of discourse network analysis (Leifeld, 2013, 2017) which brings together qualitative content analysis with social network analysis to operationalize the discourse coalitions and various aspects of it (Leifeld & Haunss, 2012). The method focuses on the discursive layer of policy-making and emphasizes policy-processes as 'networked phenomena' (Leifeld, 2017, p. 2). It is operationalized through the software called Discourse Network Analyzer (DNA), discussed in detail below.

For collecting data, three English-language newspapers with highest circulation (Audit Bureau of Circulation, 2019) were selected: The Times of India, The Hindu and The Hindustan Times. These publications represent the quality print media in India (Schmidt & Schäfer, 2015). They have considerable influence over policy-making processes of the country (Sonwalkar, 2002) and have been used extensively in the literature on Indian media discourse (Billett, 2010; Das, 2020; Mittal, 2012; Schmidt & Schäfer, 2015). The period chosen for the study marks the progress of Indian energy policy from the Paris Agreement in 2015 to the pandemic in 2020. From pledging ambitious renewable energy targets in the Paris Agreement in 2015 to integrating the expansion of coal in the Covid-19 revival plan, the energy discourse has seen conflicting developments.

The articles from the newspapers were extracted from the Factiva database employing the keywords *coal and (renewable or transition or clean energy or green energy).* Since coal is the most dominant and significant fossil fuel in the Indian context, the paper focuses on Indian energy transition specifically away from coal. It emphasizes how the transition process is being represented when discussed in the context of coal along with renewables. Therefore, the keyword search string employed consists of the terms, 'coal', 'transition' and 'renewables', with other words that are used to represent renewables. The selected keywords and their combinations have been previously used in the literature to study various aspects of energy transition(Neresini et al., 2020; Walker et al., 2019). The search yielded 406 articles. After removing the irrelevant and redundant articles, 229 articles were selected for final coding. The significant number of irrelevant articles can be attributed to the high occurrence of the keywords in the title of the Minister, Piyush Goyal, who was given multiple portfolios, namely, Power, Coal, New and Renewable Energy.

The retrieved data was transported into the software DNA. Within this method, the basic unit of analysis is the statement. This is essentially a piece of the text which is either directly quoted as being uttered by a particular actor or paraphrased. This statement is coded for four main variables:

(a) the *person* speaking the statement

(b) the *organization* which the person represents
(c) the *concept* category, which is a condensed statement summarizing what is being referred to in the statement, and

(d) a dummy variable recording the actor's *agreement/disagreement* with the concept.

For example, consider the following statement:

> The NDA government got only six months to rewrite the entire country's coal history … In the last 10 months, we have institutionalised structural changes in 84 departments … and have set up a base to ramp up the future, Goyal said.

The above statement was coded as, (a) Piyush Goyal [the person]; (b) Ministry of Coal [the organization]; Coal will stay financially viable in the future [the concept]; and agreement with the concept.

On the other hand, if we take the following statement as an example:

> Greenpeace India … questioned the economic and social justification for a coal expansion of the scale envisioned.

This statement was coded as, (a) Greenpeace [the organization]; (b) Coal will stay financially viable in the future; and (c) disagreement with the concept.

Following this procedure, we extracted 540 statements, attributed to 108 persons, 127 organizations, coded into 47 concept categories arrived at through an inductive process. Once coded the data was then imported into software Visone, through which DNA allows to visualize and analyse various networks emerging. For our study, we extracted one-mode actor and concept congruence networks the structure of which can be grasped from Figure 1.

The DNA and Visone allows for the analysis of discourse coalitions, especially the ideological affinity between actors in a discourse coalition through a 'global clustering coefficient' (Leifeld & Haunss, 2012, p. 4). The clustering coefficient used in this study is the Louvain method which

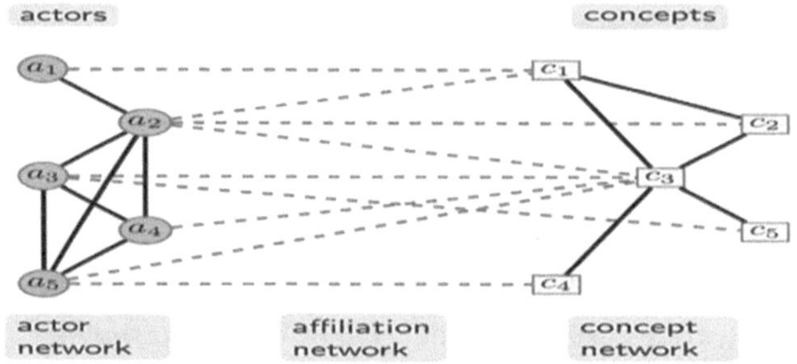

Figure 1. Basic model representing discourse networks (Leifeld, 2017, p. 307). The circular nodes represent the actors and the square ones represent the concepts. A link (dotted line) between an actor and a concept signifies that the actor has referred to that concept. A link (solid line) between actors signifies that they both have referred either positively or negatively to the same concept. Similarly, a link between concepts shows that they both have been referred to by the same actor either positively or negatively. For example, actors a1 and a2 have referred to the concept c1 in their statements, therefore connected with a link. Likewise, concepts c1 and c3 are linked since they have been referred to by actor a2.

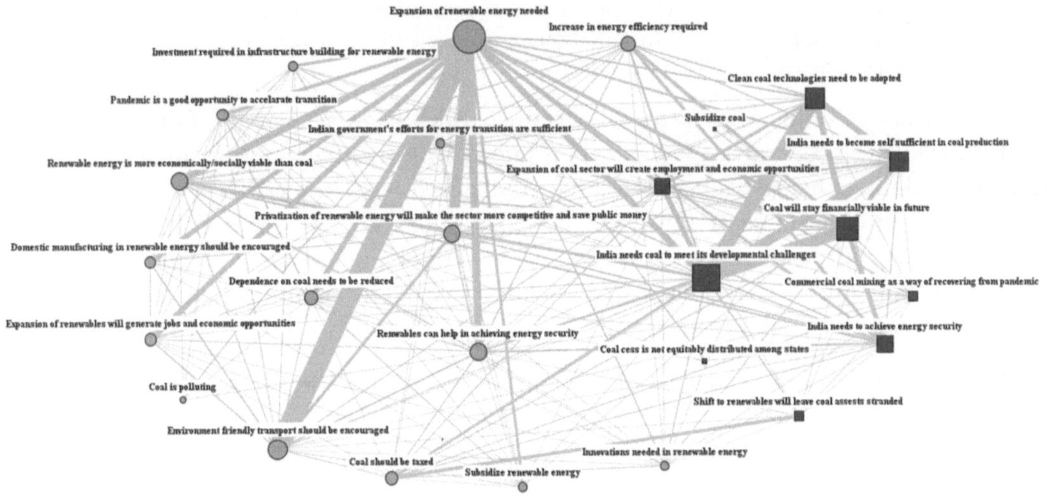

Figure 2. One mode network of the concepts. The circular nodes in green depict storylines related to the broader storyline of support for renewable energy. The square nodes in red depict the storylines related to the broader storyline of support for coal. The size of the node reflects its centrality and the width of the link depicts the frequency with which the two concepts have been referred to by the same actor. Threshold: 5 or more occurrences across all statements.

measures the strength of the links within a community against the links outside the community. It yields the coefficient value ranging from −1 to 1. A value of 0.4 or higher is usually considered to represent a meaningful clustering (Blondel et al., 2008). The method can also be used to gauge the centrality of an actor or a storyline within the discourse, the degree centrality (in %) which not only considers the frequency of occurrence of actors and storylines but also the number of times an actor is supported by other actors and the storyline is supported by the same actors. An actor in the network will be central only if it shares a greater number of ties with other actors. In terms of networks, this measure of the degree of centrality is considered a better measure than just frequencies (Kukkonen et al., 2021) Figure 2.

The dialectics within discourse: conflicting storylines, same actors

In terms of competing storylines, we broadly detected two storylines in the Indian media discourse on energy transition. The first is the pro-renewable energy storyline which subsumes a range of similar storylines that support renewable energies in various ways. Within this the most central (percentage measure given in brackets) is, 'Expansion of renewable energy needed' (17.16), followed by the 'Environment friendly transport should be encouraged' (5.88), 'Renewable energy can help achieving energy security' (4.60), 'Renewable energy is more economically/socially viable than coal' (4.39), 'Privatization of renewable energy will make the sector more competitive and save public money' (4.35). The following statement by Pujarini Sen from the NGO Greenpeace, illustrates this broad storyline:

> Given the growth in the renewable energy sector, we would have liked to see a specific target for solar and wind. Forty per cent of electricity from renewable energy by 2030 is definitely possible, especially so

> with financial and technical support… Decentralised renewable energy systems, along with grid reforms, provides an opportunity for over 300 million Indians to get access to electricity for the first time in their lives… Sen added. (Niyogi, 2015)

The second broad storyline is the pro-coal narrative, which likewise consists of different overlapping storylines. The most central among them is India's long-standing position in the national and international climate change deliberations, that: 'India needs coal to meet its developmental challenges' (11.45). In terms of centrality, it is followed by 'Coal will stay financially viable in future' (8.21), 'Clean coal technologies need to be adopted' (6.43) and 'India needs to become self-sufficient in coal production' (5.53). The wider centrality of coal storylines is reflected in the following statement from Minister of Coal, Piyush Goyal:

> I can't inject renewables into a grid that doesn't have base load. Coal will be the mainstay of our energy needs. Coal has a very bright future. (Watts, 2017)

There is a dialectical conflict between these dominant storylines, that increasingly shapes how Indian energy policy is developed and applied. There is a conflict in content, but also a strong link between these two central storylines. This reflects the divergent positions in relation to coal-fired power and renewables as reported by earlier studies (Roy & Schaffartzik, 2021); it also implies the lack of clear-cut policy to govern transition towards renewable energy (Jairaj & Kumar, 2019). The divergence and confusion are reflected in the contradictory statements made by energy minister, Piyush Goyal (Roy & Schaffartzik, 2021). As an illustration consider these two statements reportedly made by Goyal:

> Goyal states that India will do it after 150 years, when the per-capita income will be equivalent to America's and when Indians will have used up as much coal as Americans did before realizing a need for renewable resources. (Bhave, 2017)

> For us, renewable or solar is not only about economics. It's a deep passion and our PM's and nation's commitment – in that sense, all our commitment – to leave behind a cleaner world for the next generation. As the PM said when someone asked him during the US President's recent visit, 'For us there is no pressure of Lima, or Peru or Paris (global climate talks). The pressure is internal to us. It is an acquired pressure. For the future generation', Goyal told TOI on Wednesday. (Dutta, 2015)

It is notable that among the five most central storylines on energy, three are in support of coal expansion. The rhetoric of expansion for coal centres on the assumption that coal is required to meet developmental challenges and address energy poverty. This assumption drives the government's 'planned coal boom' within India (Goodman, 2016, p. 188) and is still evident in 2020, for instance in this statement by PM Modi that:

> … coal would be brought out "after decades of lockdown" and private sector involvement would help realise Rs. 33,000 crore of capital investment in the next five years. (Koshy, 2020)

Such rhetoric has significant implications since it gives opportunities to global fossil-fuel giants to capitalize and expand the coal operations (Schneider et al., 2016).

Discourse coalitions and resistance to coal expansion

To explore the avenues for resistance to coal in India, we took three storylines regarding coal which were most divisive and yielded meaningful clusters. The storylines, with the number of media statements that agreed or disagreed with them are listed in Table 1. Taking these three storylines, we discerned five discourse coalitions through our cluster analysis.

Table 1. Three most conflictual beliefs regarding future of coal.

	Storyline	Agreement (%)	Disagreement (%)	Overall frequency
1.	Coal should be taxed	82	18	17
2.	Coal will stay financially viable in future	43	57	42
3.	Commercial coal mining as a way of recovering from pandemic	50	50	06

As shown in Figure 3, the three most conflictual beliefs yielded five discourse coalitions which were intuitively grouped into two broad coalitions based on their support either for anti-coal or pro-coal storylines. The two broad coalitions clearly illustrate the conflict and tensions in Indian energy policy (Goodman, 2016). The contradiction is largely played-out between the State and non-State actors as shown in the figure. Reflecting the dominant role of state-owned CIL, the pro-coal coalition is dominated by State actors, along with some business organizations. Within it, the coalition in light grey represents a group of actors who support both the future viability of coal as well as the potential for commercial mining in pulling India out from the recent pandemic-induced economic slump. The pro-coal storyline is rooted in India's long-standing position

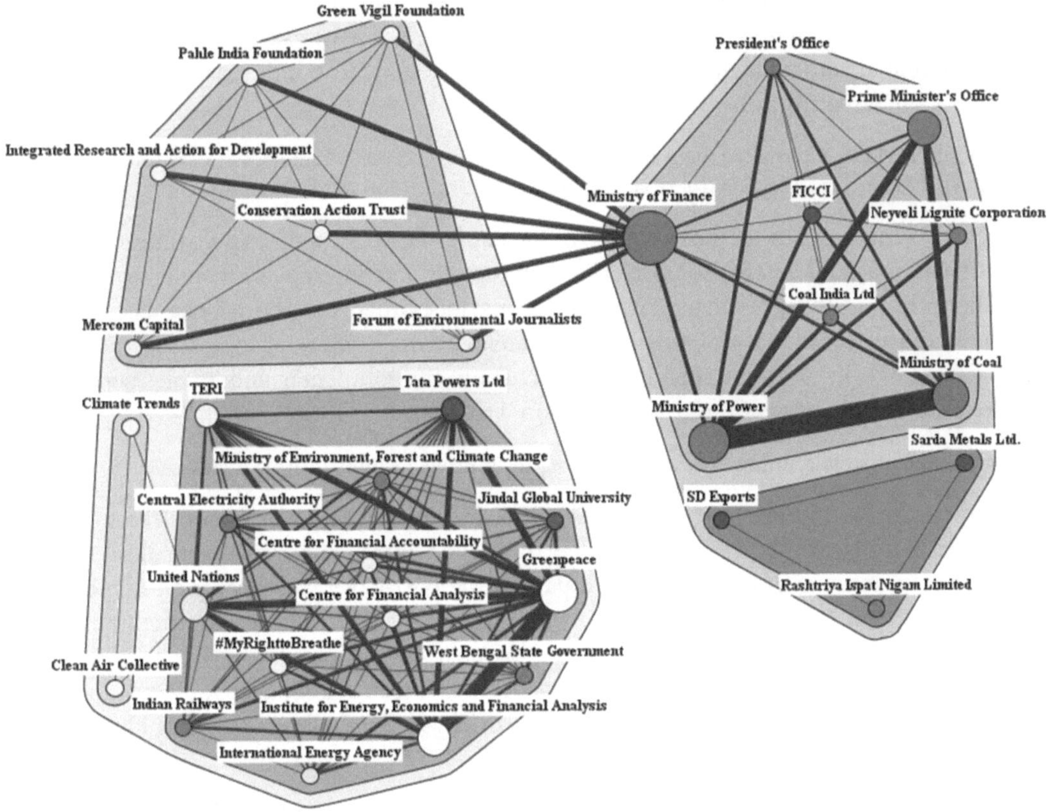

Figure 3. One mode network of organizations based on three beliefs related to coal which are most conflictual (Louvain 0.487). The nodes are colour coded to depict various types of actors (Cyan: Governmental; light blue: Inter-governmental organizations; Yellow: National Civil Society Organizations (CSOs) including think tanks and NGOs; Light yellow: international CSOs; Purple: Business organizations; Red: Research Organizations). The light green and light red meta-clusters represent two broad discourse coalitions based on anti-coal and pro-coal stances, respectively.

of looking towards coal as an answer to its various developmental and economic challenges (Lahiri-Dutt, 2014). This also helps explain why this cluster is dominated by governmental actors such as the Ministry of Finance (MoF), the MoC and the MoP. Their pro-coal stance is supported by assumptions that coal produces cheaper energy, that it offers a source of income for coal-dependent states and jobs and economic opportunities for coal dependent communities along with offering future energy security for the country. As an illustration, consider the following statement by the Indian Prime Minister, Narendra Modi in 2020, about the plan to auction 50 new coal mines:

> Mr. Modi had said the launch marked not only the implementation of reforms in the coal sector but also the creation of lakhs of jobs and would liberate the sector from decades of "lockdown". The government has decided to spend Rs. 50,000 crores on creating infrastructure for coal extraction and transport, which would also create job opportunities for locals closer to their homes, he said. (Koshy, 2020)

The centrality of governmental actors also reflects their dominant presence in energy policy making (Montrone et al., 2021). MNRE and MoC are the central policy-making bodies for renewable energy and coal, respectively. This coalition represents a coherent ideological narrative based on national sovereignty and developments which has supported coal as the 'nation-building' resource (Lahiri-Dutt, 2016) and is the main logic underlying the trend towards coal expansion.

The coalition in dark grey consists of business actors who support coal and actively resisted the proposed doubling of coal tax in the Federal budget of 2016. The reasons for their apprehension are largely economic in nature including the lack of alternatives for cheap industrial raw materials. For example, Neeraj Sarda from the Sarda Metal Ltd says:

> ... without providing alternative mechanism to steel, power, and metal sectors, who are totally dependent on coal for their production, there appears to be no justification to tax them more. (Patnaik, 2016)

The discursive challenge to the pro-coal narrative comes from the anti-coal coalition, consisting mainly of national and international civil-society actors and intergovernmental organizations. The coalition in purple represents actors who for instance broadly support a coal tax. It consists of national and international CSOs who gave statements in support of increasing the taxes on coal to discourage its use and spend the revenue on environmental and public services. For example, Debi Goenka from Conservation Action Trust said:

> The government should impose heavy tax on coal to improve health sector. (Pinjarkar, 2015)

The coalition shaded in pink brings together actors who resist the government's recent decision to open coal mining to the private sector and to auction of 50 new mines with the lifespan of 50 years. The announcement was made as a part of the post-Covid recovery plan named *Atmanirbhar Bharat*, 'self-reliant India',[4] which was released in the wake of the economic crash during the Covid-19 pandemic. The inclusion of coal in the plan was widely criticized by environmental groups such as 'Climate Trends' and the 'Clean Air Collective', a pan-India network of over 35 organizations. A statement by Aarti Khosla from Climate Trends illustrates this position:

> Coal should not be included in that plan since it will most certainly lead to increased air pollution and a host of troubles locally as well as in India's broader energy policy. (Behl, 2020)

The coalition shaded in dark green is a group of organizations who are clustered around the belief that coal is no longer economically viable and should be phased-out. It is the most diverse coalition and consists of various types of national and international actors such as CSOs (Greenpeace, TERI),

Inter-governmental organizations (United Nations, IEA), some governmental actors (Ministry of Environment, Forest and Climate Change, West Bengal State Government) and research organizations (Jindal Law University). This diversity of actors makes this coalition the most dynamic one. These actors push for a coal phase-out citing the financial viability of renewable energy against coal, and also other factors such as environmental hazards, as illustrated by the following paraphrased statement by UN Secretary-General Antonio Gutteres:

> With India facing deadly air pollution, high healthcare costs and growing disaster threats from global warming, the head of the United Nations warned the country's leaders on Friday that investing more in coal was "bad economics". (Hindustan Times, 2020)

The actors within the coalition expressed concern that, far from enabling economic recovery, coal investment and infrastructure would become an economic burden, 'stranded' as the global energy transition takes hold. Pujarini Sen from International NGO Greenpeace said:

> ... electricity from renewable energy ... will be cheaper than coal ... So where is the economic rationale for investing in what might well be stranded assets? (Niyogi, 2015)

Since the dominant narrative supporting coal is based on national sovereignty, the international non-state actors resisting coal expansion have had a conflict-ridden relationship with State actors. NGOs like Greenpeace have been cast as villains in the dominant narrative around energy, and as a consequence have faced a crackdown (Mohan & Topp, 2018). However, the continued centrality of such organizations revealed by our analysis is testimony to the constant resistance they mount against coal expansionism in India.

In addition, a range of national NGOs have been involved in a more embedded set of conflicts with the State actors. They have supported and influenced the developing discourse of the government in international climate policy while opposing the domestic state policies that go against the tenets of climate and energy justice (Swarnakar, 2019). Research NGOs like The Energy Research Institute (TERI) have played a central role in breaking India's long-standing reluctance to actively participate in climate politics. They are credited with formulating the co-benefits approach, of seeking developmental benefits from climate action, which has since become the guiding principle of climate policy-making in the country (Aamodt & Stensdal, 2017). Given this capacity to reshape the discourse, TERI's role in energy policy demonstrates the 'mere possibility' for breaking India's continued and renewed dependence on coal. Such possibility of policy co-learning, revealed in the links between the two coalitions, reflects a broad ideological congruence between state and non-state actors, especially in terms of a general consensus in favour of expanded and more inclusive economic development.

When viewed as part of a discursive network, defining a broad coalition of social forces, the individual 'capacity' (Bloch, 1986) exercised by the actors in anti-coal coalitions, can be seem as exercising a 'collective' power. While purely negative dialectical thinking (Adorno, 1973) about climate change would negate any revolutionary solution, this 'collective capacity' of discursive resistance does, we argue, present itself as 'mere possibility'. The Covid-19 pandemic has on the one hand provided a reason for coal expansion but on the on the other hand has also provided a favourable 'potentiality' for overcoming it. As such, the response to the pandemic offers the external condition for extending the scope of this 'mere possibility'. The proposal for commercial mining triggered by the economic catastrophe of the pandemic also saw widespread resistance to the decision, not just in media discourse but through actual demonstrations by local tribes and trade unions (Mazumdar et al., 2020).

Conclusion

In this study we explored the dialectical struggle unfolding in the Indian media discourse around energy transition away from coal. We extended a proposition that within the dialectics unfolding in energy systems in the context of climate change, the discursive resistance against the fossil-fuel based energy systems offers possibilities of achieving a more sustainable future. We explored the nature, content, composition and potential of such resistance through storylines, discourse coalitions and 'mere possibility'.

When viewed within the overall 'climate dialectic', India's energy policy becomes a ground for conflicting claims. Our analysis of the media narratives identified the competing storylines that constitute the overall energy discourse in India. The overall discourse is dominated by the storylines in support of renewables but among the major storylines, the ones related to coal expansion retain influence among the central actors, notably among state representatives. However, the dialectical conflict and inter-connection between the competing storylines is creating new agendas for energy policy and, we argue, offers a 'mere possibility' of ideological convergence and the emergence of a co-learning process for a sustainable future.

The unfolding of a dialectical conflict within energy discourse was most evident in our analysis of the discourse coalitions that emerged from the most conflictual storylines regarding coal. The resistance that actors in the anti-coal coalition offered to the grand narrative of the pro-coal coalition is not only discursive but has also succeeded in re-framing policies and the related discourse in the past. The ideological congruence as is evident from the high clustering score of the coalition represents a collective capacity that can be realized through more concerted efforts within the coalition to resist attempts at coal expansion.

The century-defining event of the current pandemic also carries great 'potentiality' for change on these issues. The response to the pandemic, especially in terms of the effort to enable recovery, has significantly fuelled the dialectical interaction within energy discourse in India. This is strongly evident in both storylines and discourse coalitions in India. Among the competing storylines the pandemic is cited as an opportunity to expand both renewables as well as coal. In terms of discourse coalitions, the pandemic has led to the emergence of a coalition that strongly opposes the expansion of coal and proposes alternative more effective means of overcoming economic adversities.

The implications of our findings are multifold. First, by uncovering the discursive connections among the emerging anti-coal coalitions, the study recommends more coordinated and coherent action-oriented efforts to resist coal expansion. Second, for pro-coal actors, comprising mostly of government and business actors, the study demonstrates that the legitimacy of their coal expansion strategies stands challenged by a diverse set of actors. For energy policies to be effective in striking a balance between tackling developmental and environmental concerns, policy-making needs to ensure consensus among the diverse set of actors who bring informed perspective from various vantage points. Third, our findings can promote and facilitate coordinated efforts among actors in opposing coalitions by clearly depicting the agendas that can create the common ground to bring the actors together.

There are certain limitations to the study. First, it relies exclusively on English media which limits the scope of the study in a multi-lingual country and may have missed certain key actors and storylines emerging from the rich vernacular media. Second, a limitation accompanying most studies relying on media data is that the media statements given by the actors may represent a strategic and formally derived position and may not be congruent with their actual actions and beliefs.

Multiple possibilities emerge for future research. First, it can delve deeper into the perspective of the actors who are dominant in media discourse by conducting in-depth interviews with them. Second, the future enquiries may extend examination of discursive strategies related to energy deployed in other spaces by including diverse sources of data such as policy reports and parliamentary debate. Third, more action-oriented research can be conducted that seeks to translate the discursive resistance to fossil-fuel into a more coordinated effort and also to deploy more coherent ideological devices against the grand narratives supporting the fossil-fuel regime.

Notes

1. The energy transition has been defined in multiple ways. For this paper we adopt the definition given by Araújo (2014, p. 112) where energy transition implies a 'shift in the nature or pattern of how energy is utilized within a system … (including) … fuel type, access, sourcing, delivery, reliability, or end use as well as with the overall orientation of the system.' This is a wholistic definition which cuts across multiple factors involved in the transition.
2. The concept of dialectic has a rich philosophical history and has evolved through the writings of various scholars. In this study, the concept is essentially employed in simple Hegelian terms, where the synthesizing changes emerge due to a series of conflicts between opposing forces of thesis and antithesis. As a concept as well as a method, it is useful since it not only helps to make sense of the conflicts but also opens room for examining the possibilities of resolution and reconciliation. (For a more detailed account of dialectics as a method see Sherman (1976)).
3. Storylines are different from policy core and secondary beliefs that bring together advocacy coalitions (Sabatier and Jenkins-Smith 1988). The beliefs assume a fixed position on the part of the individual which is subject to change through rational learning. Alternatively, storylines take language as constitutive of reality providing more room for change in loosely held value positions (Hajer, 1995). The fixedness inherent in the belief dependent advocacy coalitions make discourse coalitions based on storylines a better option for understanding policy processes (Fischer, 2003).
4. The plan was announced in mid-May 2020 with the stimulus of INR 20,000 crore to recover the economy from the impact of the pandemic. The rhetoric of self-reliance was invoked to harness more coal as a cheap and abundant resource available with the country. 50 coal mines were announced for commercial mining thereby promoting private players and ending the long-standing state monopoly of Coal India Ltd within the coal sector.

Disclosure statement

No potential conflict of interest was reported by the author(s).

ORCID

Rajshri Shukla http://orcid.org/0000-0002-6449-9563
Pradip Swarnakar http://orcid.org/0000-0002-9532-1215

References

Aamodt, S., & Stensdal, I. (2017). Seizing policy windows: Policy influence of climate advocacy coalitions in Brazil, China, and India, 2000–2015. *Global Environmental Change*, *46*, 114–125. https://doi.org/10.1016/j.gloenvcha.2017.08.006

Adorno, T. W. (1973). *Negative dialectics*. Seabury Press.

Ahmad, N., & Lahiri-Dutt, K. (2006). Engendering mining communities: Examining the missing gender concerns in coal mining displacement and rehabilitation in India. *Gender, Technology and Development*, *10*(3), 313–339. https://doi.org/10.1177/097185240601000302

Araújo, K. (2014). "The Emerging Field of Energy Transitions: Progress, Challenges, and Opportunities." *Energy Research & Social Science*, *1*, 112–21. https://doi.org/10.1016/j.erss.2014.03.002

Audit Bureau of Circulation. 2019. Highest circulated dailies, weeklies & magazines amongst member publications (across languages). Retrieved January 24, 2021, from http://www.auditbureau.org/files/JD%202019%20Highest%20Circulated%20(across%20languages).pdf.

Behl, M. (2020, June 23). 90% health care workers against coal mining: Survey. *The Times of India*.

Bhave, N. (2017, August 25). An Inconvenient Sequel: Truth to Power.

Billett, S. (2010). Dividing climate change: Global warming in the Indian mass media. *Climatic Change*, *99*(1–2), 1–16. https://doi.org/10.1007/s10584-009-9605-3

Bloch, E. (1986). *The principle of hope*. Vol. 1. (3rd ed). MIT Press.

Blondel, V. D., Guillaume, J.-L., Lambiotte, R., & Lefebvre, E. (2008). Fast unfolding of communities in large networks. *Journal of Statistical Mechanics: Theory and Experiment*, *2008*(10), P10008. https://doi.org/10.1088/1742-5468/2008/10/P10008

Boykoff, M. T. (2011). *Who speaks for the climate: Making sense of media reporting on climate change*. Cambridge University Press.

Bulkeley, H. (2000). Discourse coalitions and the Australian climate change policy network. *Environment and Planning C: Government and Policy*, *18*(6), 727–748. https://doi.org/10.1068/c9905j

Buschmann, P., & Oels, A. (2019). The overlooked role of discourse in breaking carbon lock-in: The Case of the German energy transition. *WIRES Climate Change*, *10*(3), e574. https://doi.org/10.1002/wcc.574

CEA. 2022. National power portal. Retrieved March 12, 2022, from https://npp.gov.in/publishedReports.

Černý, O., & Ocelík, P. (2020). Incumbents' strategies in media coverage: A case of the Czech coal policy. *Politics and Governance*, *8*(2), 272–285. https://doi.org/10.17645/pag.v8i2.2610

Das, J. (2020). The struggle for climate justice: Three Indian news media coverage of climate change. *Environmental Communication*, *14*(1), 126–140. https://doi.org/10.1080/17524032.2019.1629976

Djerf-Pierre, M., Cokley, J., & Kuchel, L. J. (2016). Framing renewable energy: A comparative study of newspapers in Australia and Sweden. *Environmental Communication*, *10*(5), 634–655. https://doi.org/10.1080/17524032.2015.1056542

Dsouza, S., & Singhal, K. (2021). *Socio-economic impacts of coal transitions in India: Bottom-up analysis of jobs in coal and coal-consuming industries*. National Foundation of India.

Dubash, N. K. (2019). *India in a warming world: Integrating climate change and development*. Oxford University Press.

Dubash, N. K., Khosla, R., Kelkar, U., & Lele, S. (2018). India and climate change: Evolving ideas and increasing policy engagement. *Annual Review of Environment and Resources*, *43*(1), 395–424. https://doi.org/10.1146/annurev-environ-102017-025809

Dutta, S. (2015, February 13). Solar plan will not skid on oil prices, Piyush Goyal says.

Fischer, F. (2003). *Reframing public policy. Discursive politics and deliberative practices*. Oxford University Press.

Ganowski, S., & Rowlands, I. H. (2020). Read all about it! comparing media discourse on energy storage in Canada and the United Kingdom in a transition era. *Energy Research & Social Science*, *70*, 101709. https://doi.org/10.1016/j.erss.2020.101709

Goodman, J. (2016). The 'climate dialectic' in energy policy: Germany and India compared. *Energy Policy, 99*, 184–193. https://doi.org/10.1016/j.enpol.2016.03.014

Gunderson, R. (2020). Dialectics facing prehistoric catastrophe: Merely possible climate change solutions. *Critical Sociology, 46*(4–5), 605–621. https://doi.org/10.1177/0896920519827350

Hajer, M. A. (1993). Discourse coalitions and the institutionalization of practice: The case of acid rain in Britain. In F. Fischer, & J. Forester (Eds.), *The argumentative turn in policy analysis and planning* (pp. 43–76). Duke University Press.

Hajer, M. A. (1995). *The politics of environmental discourse: Ecological modernization and the policy process.* Clarendon Press/Oxford University Press.

Hess, D. J. (2019). Coalitions, framing, and the politics of energy transitions: Local democracy and community choice in California. *Energy Research & Social Science, 50*, 38–50. https://doi.org/10.1016/j.erss.2018.11.013

Hindustan Times. (2020, August 29). UN chief warns India over 'bad economics' of coal.

Huber, M. T. (2009). Energizing historical materialism: Fossil fuels, space and the capitalist mode of production. *Geoforum; Journal of Physical, Human, and Regional Geosciences, 40*(1), 105–115. https://doi.org/10.1016/j.geoforum.2008.08.004

International Energy Agency. (2021). *India energy outlook 2021.*

Isoaho, K., & Markard, J. (2020). The politics of technology decline: Discursive struggles over coal phase-Out in the UK. *Review of Policy Research, 37*(3), 342–368. https://doi.org/10.1111/ropr.12370

Jairaj, B., & Kumar, P. (2019). An assessment of India's energy transition: Paris and beyond. *Jindal Global Law Review, 10*(1), 35–47. https://doi.org/10.1007/s41020-019-00085-2

Jessup, B. (2010). Plural and hybrid environmental values: A discourse analysis of the wind energy conflict in Australia and the United Kingdom. *Environmental Politics, 19*(1), 21–44. https://doi.org/10.1080/09644010903396069

Koshy, J. (2020, August 29). Guterres flays India's coal subsidy.

Kukkonen, A., Stoddart, M. C., & Ylä-Anttila, T. (2021). Actors and justifications in media debates on Arctic climate change in Finland and Canada: A network approach. *Acta Sociologica, 64*(1), 103–117. https://doi.org/10.1177/0001699319890902

Lahiri-Dutt, K. (2014). *The coal nation: Histories, ecologies and politics of coal in India.* Ashgate.

Lahiri-Dutt, K. (2016). The diverse worlds of coal in India: Energising the nation, energising livelihoods. *Energy Policy, 99*, 203–213. https://doi.org/10.1016/j.enpol.2016.05.045

Lee, H., & Birol, F. 2020. Energy is at the heart of the solution to the climate challenge. *IPCC*. Retrieved March 12, 2022, from https://www.ipcc.ch/2020/07/31/energy-climatechallenge/.

Leifeld, P. (2013). Reconceptualizing major policy change in the advocacy coalition framework: A discourse network analysis of German pension politics: Reconceptualizing major policy change in the ACF. *Policy Studies Journal, 41*(1), 169–198. https://doi.org/10.1111/psj.12007

Leifeld, P. (2017). Discourse network analysis: Policy debates as dynamic networks. In J. N. Victor, M. N. Lubell, & A. H. Montgomery (Eds.), *The Oxford handbook of political networks* (pp. 301–326). Oxford University Press.

Leifeld, P., & Haunss, S. (2012). Political discourse networks and the conflict over software patents in Europe: Political discourse networks. *European Journal of Political Research, 51*(3), 382–409. https://doi.org/10.1111/j.1475-6765.2011.02003.x

Litfin, K. (1995). Framing science: Precautionary discourse and the ozone treaties. *Millennium, 24*(2), 251–277. https://doi.org/10.1177/03058298950240020501

Mayrhofer, J., & Gupta, J. (2016). The politics of co-benefits in India's energy sector. *Environment and Planning C: Government and Policy, 34*(7), 1344–1363. https://doi.org/10.1177/0263774X15619629

Mazumdar, G., Mohanty, D., & Mishra, R. (2020, June 26). Tribals in 3 states against Centre's decision to auction coal mines, say it would displace them. *Hindustan Times.*

Ministry of External Affairs. 2021. National statement by Prime Minister Shri Narendra Modi at COP26 summit in Glasgow. Retrieved March 12, 2022, from https://www.mea.gov.in/Speeches-Statements.htm?dtl/34466/National+Statement+by+Prime+Minister+Shri+Narendra+Modi+at+COP26+Summit+in+Glasgow.

Mittal, R. (2012). Climate change coverage in Indian print media: A discourse analysis. *The International Journal of Climate Change: Impacts and Responses, 3*(2), 219–230. https://doi.org/10.18848/1835-7156/CGP/v03i02/37105

MoEFCC. (2015). India's intended nationally determined contrinution: Working towards climate justice. Retrieved May 5, 2020, from file:///C:/Users/Rajshri/AppData/Local/Temp/INDIA_INDC_TO_UNFCCC.pdf.

Mohan, A., & Topp, K. (2018). India's energy future: Contested narratives of change. *Energy Research & Social Science*, *44*, 75–82. https://doi.org/10.1016/j.erss.2018.04.040

Montrone, L., Ohlendorf, N., & Chandra, R. (2021). The political economy of coal in India – evidence from expert interviews. *Energy for Sustainable Development*, *61*, 230–240. https://doi.org/10.1016/j.esd.2021.02.003

Moore, J. W. (2015). *Capitalism in the web of life: Ecology and the accumulation of capital*. Verso.

Neresini, F., Giardullo, P., Di Buccio, E., & Cammozzo, A. (2020). Exploring socio-technical future scenarios in the media: The energy transition case in Italian daily newspapers. *Quality & Quantity*, *54*(1), 147–168. https://doi.org/10.1007/s11135-019-00947-w

Niyogi, S. (2015, October 3). Greenpeace says India's INDC targets step in right direction.

Oskarsson, P., Nielsen, K. B., Lahiri-Dutt, K., & Roy, B. (2021). India's new coal geography: Coastal transformations, imported fuel and state-business collaboration in the transition to more fossil fuel energy. *Energy Research & Social Science*, *73*, 101903. https://doi.org/10.1016/j.erss.2020.101903

Patnaik, S. (2016, March 7). Doubling clean fuel cess will hit metal sector hard, say investors.

Pinjarkar, V. (2015, February 27). Shun policy based on coal power.

Riedy, C. (2020). Discourse coalitions for sustainability transformations: Common ground and conflict beyond neoliberalism. *Current Opinion in Environmental Sustainability*, *45*, 100–112. https://doi.org/10.1016/j.cosust.2020.09.014

Rochyadi-Reetz, M., Arlt, D., Wolling, J., & Bräuer, M. (2019). Explaining the media's framing of renewable energies: An international comparison. *Frontiers in Environmental Science*, *7*, 119. https://doi.org/10.3389/fenvs.2019.00119

Roy, B., & Schaffartzik, A. (2021). Talk renewables, walk coal_ The paradox of India's energy transition. *Ecological Economics*, *180*, 106871. https://doi.org/10.1016/j.ecolecon.2020.106871

Sabatier, P. A., & Jenkins, S. (1988). An Advocacy coalition model of policy change and the role of policy oriented learning therein. *Policy Sciences*, *21*(2/3), 129–168.

Schmidt, A., & Schäfer, M. S. (2015). Constructions of climate justice in German, Indian and US media. *Climatic Change*, *133*(3), 535–549. https://doi.org/10.1007/s10584-015-1488-x

Schneider, J., Schwarze, S., Bsumek, P. K., & Peeples, J. (2016). *Under pressure*. Palgrave Macmillan UK.

Sherman, H. (1976). Dialectics as a method. *Insurgent Sociologist*, *6*(4), 57–64. https://doi.org/10.1177/089692057600600408

Sonwalkar, P. (2002). 'Murdochization' of the Indian press: From by-line to bottom-line. *Media, Culture & Society*, *24*(6), 821–834. https://doi.org/10.1177/016344370202400605

Stoddart, M. C. J., & Tindall, D. B. (2015). Canadian news media and the cultural dynamics of multilevel climate governance. *Environmental Politics*, *24*(3), 401–422. https://doi.org/10.1080/09644016.2015.1008249

Swarnakar, P. (2019). Climate change, civil society, and social movement in India. In N. Dubash (Ed.), *India in a warming world* (pp. 253–272). Oxford University Press.

Swarnakar, P., Shukla, R., & Broadbent, J. (2021). Beliefs and networks: Mapping the Indian climate policy discourse surrounding the Paris climate change conference in 2015. *Environmental Communication*, *16*(2), 145–162. https://doi.org/10.1080/17524032.2021.1973528

Talukdar, R. (2016). Hiding neoliberal coal behind the Indian poor. *Journal of Australian Political Economy* (78), 132–158.

Urry, J. (2014). The problem of energy. *Theory, Culture & Society*, *31*(5), 3–20. https://doi.org/10.1177/0263276414536747

Walker, C., Alexander, A., Doucette, M. B., Lewis, D., Neufeld, H. T., Martin, D., Masuda, J., Stefanelli, R., & Castleden, H. (2019). Are the pens working for justice? News media coverage of renewable energy involving indigenous peoples in Canada. *Energy Research & Social Science*, *57*, 101230. https://doi.org/10.1016/j.erss.2019.101230

Watts, H. (2017, July 18). Post GST, you will see prices falling: Piyush Goyal.

Democratizing global climate governance? The case of indigenous representation in the Intergovernmental Panel on Climate Change (IPCC)

Pedram Rashidi and Kristen Lyons

ABSTRACT

Indigenous communities have a particular stake in climate and energy developments, and have come to occupy a central role in both the movement toward decarbonization of industrial societies and renewable energy transition. Yet they remain underrepresented and excluded from climate policy processes. This paper critically evaluates the case of the Intergovernmental Panel on Climate Change (IPCC) to foster inclusive decision-making, with a specific focus on the place of Indigenous peoples in these processes. We demonstrate the ways in which the IPCC provides limited voice for Indigenous peoples in the Panel's deliberations. We argue the IPCC's disregard for Indigenous knowledges – and Indigenous peoples – arises from a configuration of structural problems within the governance of decarbonization and the politics of knowledge production. We conclude the IPCC faces on-going challenges in supporting the democratic functioning of climate decision-making, including in upholding Indigenous peoples' internationally recognized human rights as political actors.

Introduction

There is growing interest in the role that Indigenous peoples – and Indigenous Knowledges (hereafter 'IKs') – can play in the context of climate change. In this paper, we take a critical look at how Indigenous peoples and their laws, cultures, practices and knowledges, are approached and represented in this space. To do this, we look into reports generated by the main body of climate change research, the Intergovernmental Panel on Climate Change (IPCC). The IPCC was established in 1988, and is the only international institution that is tasked to provide a comprehensive and multidisciplinary corpus of knowledge to document, and respond to, the challenge of climate change. The IPCC reports are widely believed to be the most authoritative assessment of climate change available for policy makers, experts, and the general public (Hughes, 2015; Hughes & Paterson, 2017; Hulme & Mahony, 2010; Vasileiadou et al., 2011).

There is now some scholarly recognition of the neglect of IKs in the IPCC assessment processes, and of the limited scope for Indigenous peoples' participation in its deliberations. Scholars have shown that the Panel is biased towards the use of scientific and technical disciplines (Bjurstrom & Polk, 2011; Corbera et al., 2016; Hiramatsu et al., 2008; Hulme & Mahony, 2010), and generally neglects IKs. Ford et al. (2012; 2016a), for example, analysed Working Group II's contribution to the Fifth Assessment Report on climate change impacts, adaptation, and vulnerability, finding an underrepresentation of Indigenous expertise and limited coverage of Indigenous issues. They also identify inadequate critical engagement with the diversity and complexities of Indigenous knowledge systems. Their work points to structural issues in the politics of knowledge that have constrained Indigenous participation in the process of knowledge production. Adding to this critique, Suliman et al. (2019) demonstrate the state-centric architecture of the IPCC's global climate governance model fails to respond to the needs and interests of non-state actors, including Indigenous peoples. The outcome of this deepens Indigenous peoples' under-representation, or inclusion as 'objects' of global climate governance.

Despite this growing literature, there has been no in-depth study of the structural forces that work to limit Indigenous input into the formulation of authoritative knowledge on climate change. This paper contributes to nascent literature in this space with a diagnostic study of barriers to Indigenous engagement within the IPCC. Drawing from a content analysis of a selection of the Panel's reports, this paper demonstrates the IPCC's misrepresentation of Indigenous peoples and IKs. In so doing, we show that IKs are positioned as sources of information that may supplement science. Indigenous peoples are also situated as vulnerable to climate change, rather than actors who may be empowered by the tools of their own knowledge systems. To do this, we demonstrate that the main problem with Indigenous representation (or lack thereof) in the IPCC is not the day-to-day politics of exclusion; rather, it is the configuration of structural problems that have their roots in the history of the IPCC as a governing institution that has emerged within the context of European colonial science expansion (see Whyte, 2017). Overall, we argue that the epistemological and ideological forces shaping the structure of international knowledge institutions, such as the IPCC, allow for the inclusion of only a limited range of knowledge in their assessment processes. We conclude the IPCC must untangle these structural limits of its model of colonial global environmental governance to remedy its neglect of IKs, and invite on-going debate about pathways to reform.

Our paper begins with a brief background on relevant science, environmental governance and Indigenous Knowledges literature, which provides the context for our analysis of the IPCC. It then proceeds with the content analysis of selected IPCC reports. Based on our content analysis, we critically explore the way(s) the IPCC has come to construct IKs. Finally, we examine key underlying assumptions the IPCC draws upon in approaching Indigenous peoples and their knowledges which, we argue, results in such construction of IKs.

Science, environmental governance and Indigenous Knowledges

Our approach to the analysis of the IPCC is informed by historical legacies that shape the contemporary workings of science, science institutions and knowledge production. We therefore begin with some scene setting, locating our analysis within histories that bind the expansion of Western science with the colonial project, as well as considering the legacies of this for contemporary environmental governance, including global climate governance, the focus of this paper.

Western scientific expansion occurred alongside processes of modernization, developmentalism and colonization. The Scientific Revolution brought new ways of knowing the world, and with outcomes that sidelined, enclosed and/or co-opted, non-European knowledges. Harding (2008, p. 38) describes this demarcation project as bifurcating knowledge claims; between that valued as 'rational science', versus its 'irrational', or 'savage', 'nonscience' counterpart. This colonial project –

grounded in technocratic and developmentalist structures of domination – also enabled the expansion of Eurocentric science and scientific institutions, while at the same time sidelining Indigenous peoples (see Whyte, 2017).

Turning to contemporary environmental challenges, scientific and technical approaches remain central to both politics and governance, with scientists frequently positioned as 'experts' alongside the scientization of environmental decision making (Bäckstrand, 2004; Weingart, 1999). This is evident in the growth of science-based solutions to environmental problems. The expansion in global carbon offset initiatives, such as Reducing Emissions from Deforestation and Forest Degradation (REDD+), are exemplars of such technological-fixes that ground solutions to the challenge of climate change within a dominant modernist framework[1] (Hannigan, 1995). These development interventions position rational science-based decision making as best placed to address environmental problems. The IPCC is a product of this paradigm. Its politics of knowledge production was positioned from inception[2] as a scientific endeavour that would draw upon (mostly Northern-based) experts to generate technical – and market-based – solutions to the global[3] challenge of climate change.

The emergence of new forms of environmental and public health risks, and other kinds of uncertainties, however, have also challenged the conventional (and exclusive) role of science in decision making. Reflecting this tension, contemporary approaches to environmental problems are frequently characterized by the inclusion of a diverse array of actors as part of deliberative forms of environmental governance (see Dryzek, 2002). This participatory turn has widened the scope of what counts as knowledge, as well as who is recognized as a legitimate knowledge bearer. The proliferation of locally-driven 'bottom-up' approaches reflects the plurality of knowledges that are now recognized as part of environmental decision making (Bäckstrand, 2004).

This deliberative turn has occurred alongside growing international recognition of the rights of Indigenous peoples, including recognition of the right to self-determination and participatory rights in the global climate regime (Lindroth & Sinevaara-Niskanen, 2013), including the UN Declaration on the Rights of Indigenous Peoples (UNDRIP). Yet despite this recognition, Smith (2007, p. 200) explains 'Indigenous peoples and their counterhegemonic discourse are masked and marginalized', the outcome of which provides only limited possibilities for their voices to be heard within the IPCC.

Indigenous Knowledges can be taken to include those modes of knowledge belonging to communities, peoples, and nations that are identified with Indigeneity (Watson, 2014). For Semali and Kincheloe (1999, p. 3), IKs reflect 'the dynamic way in which the residents of an area have come to understand themselves in relationship to their natural environment and how they organize that folk knowledge of flora and fauna, cultural beliefs, and history to enhance their lives'. Agrawal (2002) avoids separating IKs from Western sciences on the basis of simple, universal criteria. Rather, he asserts that what we know today as IKs have interacted with Western knowledge since at least the fifteenth century, and on this basis – and given that both are based on systematic empirical observation – there is no strict division between Western and Indigenous knowledges. Yet IK's do have a distinctively holistic character: ultimately, as Latulippe and Klenk (2020, p. 10) assert, 'Indigenous knowledge is inextricably linked to Indigenous self-determination, rights and responsibilities, which includes respect for the obligations of all beings of creation, not only human'.

Research methodology

Our paper seeks to illuminate the political agency of Indigenous peoples in the context of global climate governance, and how Indigenous observations and responses to climate change are represented in IPCC reports. To do this, we adopt a method of content analysis. All references to Indigenous knowledges and peoples in IPCC reports have been surveyed. Given the volume of the reports, only the two latest assessment reports – i.e. the Fourth Assessment Report (IPCC, 2007a, 2007b, 2007c; hereafter 'AR4, 2007') and the Fifth Assessment Report (IPCC, 2013, 2014a, 2014b; hereafter 'AR5, 2014') – and the latest two Special Reports – i.e. 'Climate Change and Land' (IPCC, 2019a) and 'The Ocean and Cryosphere in a Changing Climate' (IPCC, 2019b) – were selected for analysis. These selected reports contain much larger numbers of references to Indigenous peoples and IKs as compared to the previous IPCC reports.[4]

To conduct the analysis, all references to the terms 'indigenous', 'aboriginal', 'native', 'indigeneity', 'indigenous knowledge', 'traditional knowledge', 'traditional ecological knowledge', and 'local knowledge' were identified. To ensure the extracted texts were representative of their respective sectional context, the paragraph, section, and chapter of those texts were systematically identified and compiled. The extracted text was then reviewed to answer the following questions:

- How are IKs approached or viewed?
- How are IKs evaluated?
- What roles, characteristics, and intellectual capacities are given to Indigenous peoples? And how are they represented and engaged with?

A summary of the analysis is shown in Table 1. The summary is accompanied by quotations from the reports that exemplify the authors' assumptions about the value of IKs for understanding and responding to climate change. Italics are added to call attention to key phrases. The full data comprising all coded and categorized extracted texts from the IPCC reports can be provided upon the request.

How the IPCC constructs 'Indigenous peoples' and 'Indigenous Knowledges'

Overall, despite the increasing recognition of the value of IKs in recent IPCC reports (e.g. IPCC, 2019a, 2019b), the analysis conducted of relevant IPCC documents – a summary of which is reported in Table 1 – demonstrates the IPCC fails to recognize IKs as knowledge systems parallel to, albeit different from, Western science. IKs are treated primarily as *sources* of *data* or *information* that can *supplement* science if they prove *useful*. This representation also prompts the question: How has the IPCC come to construct Indigenous peoples and their knowledges in such ways?

To answer this question, we begin by outlining the historical context from which the IPCC emerged. This helps elucidate the ways the Panel currently operates. We then identify several key institutional and epistemic assumptions that drive the assessment processes, thereby generating certain understandings of Indigenous peoples and IKs. We then show how the IPCC operationalizes its techno-scientific episteme through its determination of who counts as an expert and what counts as expertise.

IPCC: historical background

Discourses concerning climate change have been constructed predominantly in scientific and techno-economic terms. Environmental change has been recognized and recorded for millennia around the world, mainly at the local level. However, climate change as we know it today has exclusively developed within the Western scientific paradigm. It built on the work of the early nineteenth

Table 1. Summary of the reports analysis.

Questions	Summary of analysis	Extracted sample text from the IPCC reports
How are Indigenous Knowledges viewed in the reports?	In IPCC reports IKs are primarily assumed to be no more than *supplementary* to science, i.e. they have no status by themselves as independently authoritative worldviews or systems of knowledge. It is assumed the only way of including IKs in the dominant climate change discourse is to *incorporate* them into science and policy.	• 'The idea of usable knowledge in climate assessments [...] requires the inclusion of local knowledge, including indigenous knowledge, to *complement* more formal technical understanding generated through scientific research' (AR4-WGII: 832). • 'Indigenous knowledge can *supplement* scientific knowledge in geographic areas with a paucity of data [...]' (AR5-WGIII: 165). • '[...] recognizing and incorporating local knowledge and indigenous knowledge is increasingly emphasized in successful policy implementation' (IPCC, 2019a, p. 638).
How are IKs evaluated?	IKs in IPCC reports are often regarded as merely important *sources* of *information* or *data*, rather than as independent knowledge systems. The reports use 'indigenous knowledge' in the singular form as if all are essentially the same, and the variations are inconsequential, and glossing over the actual diversity of the knowledges of Indigenous peoples.	• 'There is recent recognition of the *untapped resource* of Indigenous knowledge about past climate change which could be used to inform adaptation options' (AR4-WGII: 523). • 'Indigenous knowledge is an important *resource* in climate risk management and is important for food security in many parts of the world' (AR5-WGII: 520). • '[...] actions are most successful when [...] using local knowledge and Indigenous knowledge' (IPCC, 2019b, p. 30).
What roles, characteristics, and intellectual capacities are given to Indigenous peoples? And how are these groups represented and engaged with?	In the reports indigenous peoples are generally assumed to be passively *vulnerable* to climate change and incapable by themselves of adapting to it. They are thus assumed to be in need of help from the 'developed world'.	• 'Special attention needs to be given to indigenous peoples with subsistence livelihoods and groups with limited access to information and few means of adaptation' (AR4-WGII: 248). • 'most vulnerable to climate change including indigenous peoples and local communities, women, and the poor and marginalised' (IPCC, 2019a, p. 28) • 'Some valuable experience has been gained recently on [...] involvement of *vulnerable* or marginalized groups such as indigenous peoples ... ' (AR5-WGII: 1227).

century natural philosophers who pioneered the scientific investigation of the dynamics of the Earth's climate (Hulme, 2009) and later evolved with the help of computer modeling in the second half of the twentieth century. Furthermore, as Cox (1996) argues, from the 1970s the world economy began to transform, shifting the power nexus for environmental politics. Multinational corporations and international financial institutions emerged as the primary agents of economic development, while states took on the role of attracting global investors to advance their development priorities. The rise of non-state actors was largely limited to powerful economic agents (i.e. multinational corporations, international organizations, individual entrepreneurs, etc.) while others, including Indigenous groups, remained on the margins, both domestically and in international forums.

Within that context, global environmental governance was largely shaped by international epistemic communities that promoted global-scale science (Miller, 2004) and transnational economic forces that pursued economic growth to maintain the existing liberal global order (Bernstein & Pauly, 2012). In this highly politicized environment, the IPCC was established as an *inter-governmental* panel by states and tasked to provide a clear scientific view on climate change and its potential environmental and socio-economic impacts. From the outset, the Panel positioned itself as an inter-state scientific body with a rigorous peer-review process that could only be applied to (modern Western) scientific research. Consequently, other forms of knowledge were left in the categories of non-scientific information. The embedded 'scientism' in the IPCC has emerged through historical processes in which Western understandings of climate change and institutions that have defined it are co-produced (Miller, 2004). This history demonstrates a notable lack of discussion of colonialism and its impacts on both Indigenous peoples and anthropocentric climate change discourses (Whyte, 2017).

IKs as supplementary to science and 'useful' sources of information

As shown in the report analysis, the IPCC views IKs as supplementary to science, meaning their inclusion as 'knowledge' is possible only through their integration into scientific literature. Our analysis also demonstrates that the IPCC regards IKs as under-used resources (AR4-WGII: 139) which could be valuable to complement the work of scientists and planners (AR4-WGII: 456), especially to increase the effectiveness of climate change adaptation programmes (AR5-WGII: 26). The utility of IKs is therefore contingent on their suitability for calibration as scientific knowledge, thereby improving the efficacy of policy models (AR5-WGII: 844). The Panel goes as far as viewing the whole bio-cultural heritage of Indigenous peoples as 'an irreplaceable bundle of teaching on the practices of mitigation and sustainability' (AR5-WGIII: 255).

To better understand how the IPCC has come to view Indigenous peoples and their knowledges in such an explicitly utilitarian way, one needs to appreciate the broader historical-scientific context. This valuing of IKs only in relation to scientific knowledge demonstrates legacies of the demarcation project that occurred as part of the early spread of European science (see Harding, 2008).

IKs have long been approached in an instrumental and utilitarian manner and evaluated and defined on the basis of their usefulness for scientific and economic practices. To facilitate their usability, IKs have been conceptualized as unitary knowledge systems that are shared equally across all members of any community. That essentialism has generated a notion of IKs as 'given' and 'pristine' knowledge that is 'simply waiting to be tapped into' (Briggs, 2005, p. 107). Given this understanding, it is not surprising there is an increased appetite on the part of academics, government agencies, international organizations, and corporations for access to 'reservoirs' of IKs that are potentially useful for environmental conservation (cf. Verran, 2002).

The epistemological forces that press towards the integration of IKs into science reveals a tendency to treat IKs as an 'intellectual product' that can be isolated from its social context so as to make it a useful addition to the scientific corpus (Nadasdy, 1999). This process of scientization, which reproduces IKs through the logic of development and the epistemic filter of science, creates fragments of IKs that are able to be used as supplements to science (Agrawal, 2002). Such a utilitarian approach is not only damaging and exploitative, it also reconstructs IKs in objective and abstract ways that break linkages with their place of origin and their populations.

There are at least two consequences of this reductionist view of IKs: First, the utilitarian view denies the status of IKs as independent authoritative knowledge systems and reduces them to data and information to be used by scientists. From this perspective, climate change can only be imagined from the perspective of Western science. Yet that construction ignores the strong historical evidence that many Indigenous peoples have systematically observed, recorded, and responded to such changes from their own perspectives (Berkes, 2017, pp. 171–192; Nunn & Reid, 2016; Nyong et al., 2007).

Second, such a utilitarian approach to IKs has been a major stimulus for bioprospecting and rent-seeking activities aimed at accessing economically valuable ecological knowledge held by Indigenous communities. This has opened up 'new frontiers for commercial exploitation' of both Indigenous peoples and their habitats (Schroeder, 1999, p. 2). The widely used Clean Development Mechanism of the UNFCCC, a project that aims to appropriate the carbon sink capacity of many Indigenous peoples' lands to offset industrial emissions of the global north; and the 'green developmentalism' (McAfee, 1999) of the World Bank in the Mekong Dam project that led to unprecedented socioecological damage (Goldman, 2005), each exemplify global regimes that prepare the way for utilitarian and exploitative moves on Indigenous peoples' resources. The workings of the IPCC can be seen as part of this new eco-economic mode of governance.

Indigenous peoples as vulnerable populations

The construction of IKs in IPCC reporting as databases of potentially useful information defines those knowledges as archives, and as therefore static, and belonging to the past. In so doing, Indigenous peoples are rendered unreservedly vulnerable, and as unable to adapt to a changing climate. This concept of vulnerability to climate change is a key theme in the IPCC assessment reports, where Indigenous peoples are generally assumed to be the most vulnerable amongst all populations to experience the worst climate change. They are thus deemed as requiring special attention from developed countries. Entangled with that is the Panel's focus on how Indigenous communities are isolated by their low rates of literacy (AR5-WGII: 630). Sometimes they are even put in the same category as the poor, children or disabled persons (AR5-WGII: 1227; IPCC, 2019a, p. 28).

It is undeniable that climate change is posing an unprecedented challenge to Indigenous peoples across the world; however, presuming them to be simply vulnerable not only contradicts historical facts but devalues their collective knowledge. Indigenous peoples survived many cycles of natural environmental upheaval and climate change for millennia before European colonialism began (see discussions on Pacific Islands in Campbell, 2009, and on African Sahel in Nyong et al., 2007). The presumption of vulnerability also reinforces a false association of oral cultures with illiteracy, ignoring the ways many oral Indigenous traditions carry centuries, if not millennia, of knowledges and experiences (see for example, Nunn & Reid, 2016).

Vulnerability is thus exposed as a social construct that, in the IPCC reports, is implicitly defined according to measures of development. The extent to which Indigenous peoples are assumed to be vulnerable is as much subjective as factual. Thus, anticipated sea level rise in the current century is deemed threatening for the Maldives, Tuvalu, and Kiribati. Yet, how well those vulnerable nations may be able to respond to climate change impacts is highly contested (see Sulamin et al., 2019). Ross Garnaut, a prominent economist who led the Australian government's review on climate change in 2008, proposed a solution for inundated island countries in the form of mass migration.[5] But for many affected populations, leaving their ancestral lands is not an option. They believe they will survive this crisis as they have overcome past naturally caused climatic changes (Mortreux &

Barnett, 2009). For example, Petheram et al. (2010) found that for the Yolngu people of North East Arnhem Land in Australia, climate adaptation practices were not limited to mass migration and proposed mechanical solutions. To them, staying strong; maintaining cultural, physical, and emotional health; passing down knowledge and tradition as a means for survival; close observation of the environment; and kinship support and practices of sharing are as important.

It is evident that framing climate adaptation as 'vulnerability-reduction' affects how climate change is known and addressed by and for Indigenous peoples. What seems to dominate scientifically based socioeconomic responses to climate change is a linear model of impacts and adaptation. Climate is treated as an external system that generates threats; societies are social structures that are unevenly vulnerable to climatic impacts. Possibilities for adaptation around the world thus can be captured in a universal schematic of degrees of climatic threat and units of vulnerability. This linear cause–effect-response model, from one perspective, serves the technocratic elites who seek to contain climatic threats within existing institutional frameworks (Taylor, 2015, xii, 32). From another perspective, it is able to depict many parts of the world as dangerous (disaster-prone and poverty-stricken), with inhabitants who are culturally inferior, incapable, and mere victims (Bankoff, 2001).

The symptoms of climate vulnerability might be predominantly physical (e.g. the physical effects of sea level rise). However, the underlying causes of vulnerability are economic, social, and political, and many of these causes are embedded in adaptation and development programmes (Cardona, 2004). In the case of the Canadian Arctic, Haalboom and Natcher (2012) report that the Inuit's displacement by the government was rationalized by their inability to rely any longer on wildlife resources. They were thus urged to resettle and adopt an industrial wage-earning lifestyle. When they did, the Inuit not only lost control over their land and resources but were represented as vulnerable to social and ecological change and became dependent on government support. In a similar way, the IPCC's construction of Indigenous peoples as vulnerable may create a justification for further development interventions.

Underlying assumptions in the institutional design and practices of the IPCC

Since its establishment, the IPCC's commitment to scientific practice on the one hand, and increasing attention to IKs on the other, has driven it toward further recognition of IKs. However, as our content analysis illustrated, such recognition has occurred from a particular standpoint, wherein IKs were approached in a utilitarian way and Indigenous peoples were represented in ways that lack agency – (thereby resonating with familiar colonial narratives that portray them as passive, vulnerable, victim, etc.). In this section, we now examine key underlying elements the Panel draws upon as it operationalizes its approach to Indigenous peoples and IKs. Drawing from prior research and our analysis, we demonstrate key ideologically-loaded institutional assumptions and practices that ground its assessments: consensus-building, policy relevance, and the construction of climate experts and expertise.

Consensus-building

The universal rationality of science requires that the same rules and standards apply to all scientific practices; this, in turn, facilitates epistemic consensus (Kuhn, 1962, p. 11). This looks attractive from a policy perspective, as broader consensus, supposedly, promises greater public trust (Beck et al., 2014) and political legitimacy (Dryzek & Niemeyer, 2006). The IPCC has been open and explicit about its commitment to scientific practice and its aim to generate a scientific consensus around climate change (Hulme & Mahony, 2010).[6],[7] In fact, from the beginning, one goal for establishing such an institution was to build consensus for political action within the international community (Bodansky, 1992, p. 63).

The logic of consensus-building utilized in the IPCC assessments ignores the fact that reaching epistemic consensus between knowledge claims from science and IKs – without 'translating' one to the other – seems extremely difficult, if not impossible. This difficulty lies in the fundamental differences between most Indigenous knowledge systems and Western modern science: one is oral, the other is written; one is place-based, the other is universal; one is monist, the other is dualist, and so on (see Aikenhead & Ogawa, 2007). The history of colonialism presents how the West has consistently, from a positional superiority, translated others' cultures and knowledges (including those of Indigenous peoples) through the filter of science (Chakrabarty, 2000, pp. 3–23; Said, 2003, p. 7). This has resulted in a process whereby one side (the West), appropriates and exploits the other side's (Indigenous peoples') cultures and resources.

In response to such compulsion for consensus-building, many have argued in favour of dissent as part of a more pluralistic and deliberative epistemic practice (Longino, 2002). Solomon (2006) suggests that a situation of dissent can offer more informative outcomes and produce better decisions than narrowly consensually-based ones, provided that all competing knowledge claims have shown empirical success. She notes that dissenting positions can often produce insights that could be lost via consensus building. For example, the Kogies of Colombia have a system of allocating permits for planting specific crops according to lineage affiliation and the revelations of divination (*sewa*) (Reichel-Dolmatoff, 1982). This cannot be rendered into knowledge recognizable by modern agricultural science and any attempt to construe it as knowledge supplementary to science would strip the Kogies' agricultural practice of any insight it might have for other Indigenous groups living in similar ecologies and cultures.

Similarly, scholars who have studied the IPCC have pointed to the consensus approach in the Panel as a weakness, rather than strength (Hulme, 2010; Jasanoff, 2010; Van der Sluijs et al., 2010). They argue that the IPCC has not only excluded a diversity of perspectives, differences, and ambiguities on climate change; it has also, unwittingly, concealed both scientific and political dissent, further politicizing climate science (Hulme, 2010; Jasanoff, 2010; Van der Sluijs et al., 2010). Furthermore, given the existing global power structure, a consensus-based approach in practice leads to imposing worldviews, ideologies, and epistemologies drawn from the global North onto the South (Turnhout et al., 2016). In this context, Ford et al. (2016a) referred to the emphasis on consensus in the IPCC as a homogenizing force that is insensitive to diverse cultures and ways of knowing, including those of Indigenous peoples. Making a sort of consensual science that aspires to perfect universality weakens understanding of many aspects of climate change impact and response potential that are place- and culture-specific.

Policy-relevance

To assert their political neutrality and thereby gain legitimacy, knowledge institutions such as the IPCC claim their work is 'policy-relevant' and never 'policy-prescriptive'. Contemporary scholars of science studies have challenged this claim by arguing that science and policy are co-produced, and thus, such a separation is artificial (Jasanoff, 2004; Miller, 2004). A further and more pressing issue here is whose policies the work of the IPCC is intended to be relevant for. This needs to be understood in view of the overall directives of the IPCC.

From a scientific perspective, the IPCC is tasked with assisting global climate governance by researching and recommending the best ways to manage global mean temperature. From the perspective of the dominant policy-making paradigm, which is set to employ rational and techno-economic means of analysis (Miller & Robbins, 2007, pp. 465–479), the IPCC's climate policy recommendations must be techno-economic. The link between the two perspectives is well established and appears as the backbone of global climate governance: To curb temperature rises, global GHG emissions must be contained below a certain limit. To operationalize that, GHG emissions targets must be incorporated in an economic system, so the self-regulation and efficiency inherent in the market, as well as technological innovation, contain emissions under the specified safe maximum level. The Kyoto Protocol's flexibility mechanisms, including international offsetting and the CDM, are essentially built on this premise. In establishing such a techno-economic system of governance, the areas of expertise deemed relevant have proliferated to include, for example, all subfields of environmental economics and finance. Knowledge institutions such as the IPCC have therefore turned to experts in those disciplines to produce their assessment reports.

Policy relevance is necessarily selective. Turnhout et al. (2016, p. 67) note that 'the construction of policy relevant knowledge is a political act that involves choices about the preferred audiences of knowledge and the types of policy actions that may follow from this knowledge'. Thus, the IPCC's commitment to policy-relevance has the effect of reasserting the existing epistemic supremacy of techno-science, rendering Indigenous voices and concerns unheard. This arises out of a configuration of structural problems that have their roots in the inability of technocratic institutions to accommodate diverse perspectives and worldviews. The policy relevance imperative for international institutions such as the IPCC facilitates such exclusion.

Researchers engaged directly with Indigenous communities provide numerous examples of mismatches between their understandings and the knowledge frameworks employed by the IPCC. Boillat and Berkes (2013), for example, report that Indigenous Quechua farmers in Bolivia think the underlying causes of climate change are the bad things people do and their lack of respect for Quechua customs. Similarly, according to Tschakert (2007), villagers of the Old Peanut Basin in Senegal see the loss of tradition, the lack of respect for nature, and deforestation as the key causes of climate change. Villagers of Eastern Tibet, who traditionally understood weather entirely as a local phenomenon governed by local deities, attribute recent severe weather conditions and environmental changes to the deities' anger at their 'neglect of religious duties or breach of taboos' (Byg & Salick, 2009, p. 165).

Central to these claims is the contention that it is change in human values and behaviour toward nature that requires attention. Temperature rise is only a symptom of this greater problem for many of those communities. Yet, current policy paradigms are structurally dismissive of such Indigenous perspectives unless they are translatable to scientific facts and proved to be useful supplements that support associated policies. The institutions that attempt to make knowledge about climate change relevant to existing policies only reinforce the systemic exclusion.

Expert and expertise

The composition of IPCC authors also reflects the Panel's techno-scientific epistemic premises. In selecting authors, the IPCC determines who counts as an expert and what counts as expertise. Corbera et al. (2016) demonstrated the dominance of scientific and economic disciplines in IPCC Working Group III (AR5), with authors predominantly from the Global North. Similarly, the analysis of Ford et al. (2012) showed that only nine out of 309 authors of IPCC Working Group

II (AR5) had publishing records in climate change and Indigenous studies. However, such scientization of expertise obscures the strongly political nature of the IPCC and its selection processes. As Wynne and Lynch (2015) argue, the credibility of scientists as experts goes beyond the boundaries of science, as they are treated also as authorities on questions that have cultural, ethical, political, and legal implications.

Miller (2004, p. 47) contends that the Panel has constructed a discourse that frames climate change as a global risk, and, to manage this risk, 'offered a model of global politics in which experts and expert knowledge, as politically neutral agents, were accorded significant power to define problems of global policy'. The result is a circular dynamic in which the IPCC exerts power through its assessment reports by legitimizing certain global policy institutions. In return, 'the articulation of what counts as good science will depend heavily on political institutions for support and legitimacy' (Miller, 2004, p.65).

The definition of expertise used by the IPCC, then, cannot be seen merely as a technical task. It is also a political practice that excludes those holding different worldviews. Filer (2009) describes this political practice of exclusion in the production of the Millennium Ecosystem Assessment (MA). He notes that the MA conceptual framework may at first appear to be open to diverse ways of knowing insofar as it includes cultural and religious advocacy that could affect the consumption of ecosystem services. However, the institutional necessity for establishing 'a global form of expert consensus within a specific set of environmental policy regimes' leaves no space for the inclusion of others who are not considered as experts (Filer, 2009, pp.87–88). An exclusive epistemic community in the MA comprised scientists, environmental economists and utilitarian ecologists, similar to that found in the IPCC (Carey et al., 2014; Corbera et al., 2016), has led to the oversimplification of social and cultural complexities (Filer, 2009). Within the epistemic settings of international institutions such as the UNEP, the International Council for Science, and the IPCC, neither can Indigenous knowledge holders be regarded as experts, nor can IKs meet the criteria for bona fide sources of expert knowledge.

Conclusions

This paper has demonstrated that the IPCC is structurally resistant to drawing upon non-Western and non-science-based knowledge systems, more specifically, those of Indigenous peoples. Reporting on a qualitative content analysis of selected Panel's reports, we have demonstrated that IKs are treated by the Panel as supplementary to science-based knowledge, and are addressed only insofar as they may be seen as sources of additional data. The IPCC's commitments to consensus-based, policy-relevant, and expert-led knowledge production render IKs without status as independently authoritative knowledges. IKs are only deemed relevant in relation to their capacity to bolster Western scientific knowledge. This whole process, therefore, appears to implicitly disempower Indigenous peoples. The chapter outlines of the Sixth Assessment Report that is due to be released in 2021–2022, do not indicate a major shift in this regard.

The instrumental engagement with IKs is a symptom of a deeper structural problem that extends beyond the work of the IPCC. Scholars who have studied other international environmental organizations and regimes demonstrate similar patterns in the exclusion of Indigenous voices. These include, for example the UNFCCC (Ford et al., 2016b; Shawoo & Thornton, 2019); the Kyoto Protocol (Smith, 2007); the Millennium Ecosystem Assessment (Filer, 2009); and the Intergovernmental Science-Policy Platform on Biodiversity and Ecosystem Services (IPBES) (Turnhout et al., 2014). Similarly, these organizations have evidenced considerable inertia in creating mechanisms to alter their assessment processes in ways that might facilitate genuine engagement with Indigenous peoples. This resistance occurs against a backdrop of wider progress in international engagement with Indigenous peoples, as seen within the United Nations Permanent Forum on Indigenous Issues (UNPFII) and recognition of Indigenous struggles, such as through the adoption of the 2007 United Nations Declaration on the Rights of Indigenous Peoples (UNDRIP). Despite these – and other – international advancements in the legal recognition of Indigenous peoples' rights, the entanglement of agendas for economic development across global climate governance with colonial legacies sets the scene for an extractivist relationship between the IPCC and Indigenous peoples.

The IPCC has come a long way from its first publication in 1990 with only a handful of general references to Indigenous peoples, to its recent publications (IPCC, 2019a, 2019b), wherein the Panel acknowledged the limitations of Indigenous involvement in IPCC assessments (IPCC, 2019b, p. 263) and suggested knowledge co-production for including IKs (IPCC, 2019b, pp. 103–105). However, the structural barriers that are, for instance, driven by the IPCC's strict scientific standards – a source of credibility for the Panel – treats IKs as bundles of information and depoliticizes Indigenous engagement with science. For example, in many instances, the Panel positioned Indigenous peoples as one group of 'stakeholders' next to private sector actors (IPCC, 2019a, p. 62). This ignores centuries of (neo) colonialism that entailed all sorts of deliberate attempts to assure Indigenous peoples not hold any agency and never gain self-determination. Within such a historical context, there is a legitimate concern about the contemporary 'Indigenous turn' to be another form of exploitation to unfairly share the burden of responsibility for ecological disaster with Indigenous peoples. Such attempts are not only problematic on ethical grounds, but also question the feasibility of dominant policies for a democratic transition to a low carbon society that requires fundamental shifts from existing economic and development activities.

Meaningful institutional modifications to the IPCC that facilitate equal exchanges and build trust between Indigenous knowledge holders and non-Indigenous experts may be more effective. More research will be required to understand the dynamics and effectiveness of such reforms. It is also clear from our analysis that more foundational work is required to enrich existing environmental research. International organizations, such as the IPCC, must find ways to deploy more innovative knowledge paradigms and research methods as part of their work. While this paper has focused on diagnosing the structural barriers to Indigenous engagement in the IPCC, it also urges debate about the pathways to reform.

Meantime, we should not lose sights of the many ways in which Indigenous peoples worldwide are engaging in action for climate justice. Their activities sit at, and go beyond, the limits of formal climate governance mechanisms. Working in this way, Indigenous peoples are not just pointing to the alternatives to reform within international organizations. More fundamentally, they are asserting their inalienable rights to respond to colonially-induced environmental change, and their agency in driving self-determined responses.

Notes

1. The threat of the REDD+ scheme unified and mobilized Indigenous peoples' on climate policy like never before – providing a powerful example of the Indigenous climate justice movement – with outcomes that effectively prevented its full inclusion as a UN recognized offset (Long et al., 2010). More broadly, REDD+ provided a unified platform for a wide range of justice-based resistance movements, and ground in Indigenous rights and self-determination (Suiseeya, 2017).

2. The IPCC was established in 1988 to provide assessments that influence both the UNFCCC and Kyoto Protocol, neither of which mention Indigenous peoples (see Suliman et al., 2019).
3. We note here the affect of this 'global' discourse in silencing and/or obfuscating power relations, as well as privileging state interests. As an organizing principle for international negotiations, the 'global' narrative also has a significant impact in rendering invisible the local and Indigenous peoples, a theme we return to in our analysis below.
4. For example, a simple word search in the IPCC's First and Second Assessment Reports shows only a handful of references to Indigenous-related issues, in comparison to hundreds of references to the same issues in AR4, AR5, and IPCC (2019a, 2019b).
5. http://www.radioaustralia.net.au/international/radio/onairhighlights/ross-garnaut-writes-off-pacific-states-on-climate-change.
6. http://www.ipcc.ch/organization/organization.shtml.
7. https://www.ipcc.ch/pdf/supporting-material/EM_Potential_Studies_web.pdf.

Disclosure statement

No potential conflict of interest was reported by the author(s).

Funding

The author(s) reported there is no funding associated with the work featured in this article.

References

Agrawal, A. (2002). Indigenous knowledge and the politics of classification. *International Social Science Journal*, *54*(3), 287–297. https://doi.org/10.1111/1468-2451.00382

Aikenhead, G., & Ogawa, M. (2007). Indigenous knowledge and science revisited. *Cultural Studies of Science Education*, *2*(3), 539–620. https://doi.org/10.1007/s11422-007-9067-8

Bäckstrand, K. (2004). Scientisation vs. civic expertise in environmental governance: Eco-feminist, eco-modern and post-modern responses. *Environmental Politics*, *13*(4), 695–714. https://doi.org/10.1080/0964401042000274322

Bankoff, G. (2001). Rendering the world unsafe: 'vulnerability' as western discourse. *Disasters*, *25*(1), 19–35. https://doi.org/10.1111/1467-7717.00159

Beck, S., Borie, M., Chilvers, J., Esguerra, A., Heubach, K., Hulme, M., Lidskog, R., Lövbrand, E., Marquard, E., Miller, C., Nadim, T., Neßhöver, C., Settele, J., Turnhout, E., Vasileiadou, E., & Görg, C. (2014). Towards a reflexive turn in the governance of global environmental expertise. The cases of the IPCC and the IPBES. *GAIA - Ecological Perspectives for Science and Society*, *23*(2), 80–87. https://doi.org/10.14512/gaia.23.2.4

Berkes, F. (2017). *Sacred ecology* (4th ed.). Routledge.

Bernstein, S., & Pauly, L. (2012). *Global liberalism and political order: Toward a new grand compromise?* State University of New York Press.

Bjurstrom, A., & Polk, M. (2011). Physical and economic bias in climate change research: A scientometric study of IPCC third assessment report. *Climatic Change*, *108*(1-2), 1–22. https://doi.org/10.1007/s10584-011-0018-8

Bodansky, D. (1992). Managing climate change. *Yearbook of International Environmental Law*, *3*(1), 60–74. https://doi.org/10.1093/yiel/3.1.60

Boillat, S., & Berkes, F. (2013). Perception and interpretation of climate change among Quechua farmers of Bolivia: Indigenous knowledge as a resource for adaptive capacity. *Ecology and Society*, *18*(4). https://doi.org/10.5751/ES-05894-180421

Briggs, J. (2005). The use of indigenous knowledge in development: Problems and challenges. *Progress in Development Studies, 5*(2), 99–114. https://doi.org/10.1191/1464993405ps105oa

Byg, A., & Salick, J. (2009). Local perspectives on a global phenomenon—climate change in Eastern Tibetan villages. *Global Environmental Change, 19*(2), 156–166. https://doi.org/10.1016/j.gloenvcha.2009.01.010

Campbell, J. (2009). Islandness: Vulnerability and resilience in Oceania. *Shima: The International Journal of Research Into Island Cultures, 3*(1), 85–97.

Cardona, O. D. (2004). The need for rethinking the concepts of vulnerability and risk from a holistic perspective: A necessary review and criticism for effective risk management. In G. Bankoff, G. Frerks, & D. Hilhorst (Eds.), *Mapping vulnerability: Disasters, development and people* (pp. 56–70). Routledge.

Carey, M., James, L., & Fuller, H. (2014). A new social contract for the IPCC. *Nature Climate Change, 4*(12), 1038–1039. https://doi.org/10.1038/nclimate2442

Chakrabarty, D. (2000). *Provincializing Europe: Postcolonial thought and historical difference.* Princeton University Press.

Corbera, E., Calvet-Mir, L., Hughes, H., & Paterson, M. (2016). Patterns of authorship in the IPCC Working Group III report. *Nature Climate Change, 6*(1), 94–99. https://doi.org/10.1038/nclimate2782

Cox, R. (1996). A perspective on globalization. In J. Mittelman (Ed.), *Globalization: Critical reflections* (pp. 21–30). Lynne Rienner Publishers.

Dryzek, J. (2002). *Deliberative democracy and beyond: Liberals, critics, contestations.* Oxford University Press.

Dryzek, J., & Niemeyer, S. (2006). Reconciling pluralism and consensus as political ideals. *American Journal of Political Science, 50*(3), 634–649. https://doi.org/10.1111/j.1540-5907.2006.00206.x

Filer, C. (2009). A bridge too far: The knowledge problem in the millennium assessment. In J. Carrier & P. West (Eds.), *Virtualism, governance and practice: Vision and execution in environmental conservation* (pp. 84–111). Berghahn Books.

Ford, J., Maillet, M., Pouliot, V., Meredith, T., Cavanaugh, A., and IHACC Research Team. (2016b). Adaptation and indigenous peoples in the United Nations framework convention on climate change. *Climatic Change, 139*(3-4), 429–443. https://doi.org/10.1007/s10584-016-1820-0

Ford, J., Vanderbilt, W., & Berrang-Ford, L. (2012). Authorship in IPCC AR5 and its implications for content: Climate change and Indigenous populations in WGII. *Climatic Change, 113*(2), 201–213. https://doi.org/10.1007/s10584-011-0350-z

Ford, J. D., Cameron, L., Rubis, J., Maillet, M., Nakashima, D., Willox, A. C., & Pearce, T. (2016a). Including indigenous knowledge and experience in IPCC assessment reports. *Nature Climate Change, 6*(4), 349–353. https://doi.org/10.1038/nclimate2954

Goldman, M. (2005). *Imperial nature: The World Bank and struggles for justice in the age of globalization.* Yale University Press.

Haalboom, B., & Natcher, D. (2012). The power and peril of "vulnerability": approaching community labels with caution in climate change research. *Arctic, 2012*, 319–327. http://doi.org/10.14430/arctic4219

Hannigan, J. (1995). The postmodern city: A new urbanization? *Current Sociology, 43*(1), 151–217. https://doi.org/10.1177/001139295043001011

Harding, S. (2008). *Sciences from below: Feminisms, postcolonialities, and modernities.* Duke University Press.

Hiramatsu, A., Mimura, N., & Sumi, A. (2008). A mapping of global warming research based on IPCC AR4. *Sustainability Science, 3*(2), 201–213. https://doi.org/10.1007/s11625-008-0058-9

Hughes, H. (2015). Bourdieu and the IPCC's symbolic power. *Global Environmental Politics, 15*(4), 85–104. https://doi.org/10.1162/GLEP_a_00323

Hughes, H., & Paterson, M. (2017). Narrowing the climate field: The symbolic power of authors in the IPCC's assessment of mitigation. *Review of Policy Research, 34*(6), 744–766. https://doi.org/10.1111/ropr.12255

Hulme, M. (2009). *Why we disagree about climate change? Understanding controversy, inaction and opportunity.* Cambridge University Press.

Hulme, M. (2010). Problems with making and governing global kinds of knowledge. *Global Environmental Change, 20*(4), 558–564. https://doi.org/10.1016/j.gloenvcha.2010.07.005

Hulme, M., & Mahony, M. (2010). Climate change: What do we know about the IPCC? *Progress in Physical Geography: Earth and Environment, 34*(5), 705–718. https://doi.org/10.1177/0309133310373719

IPCC. (2007a). *Climate change 2007: The physical science basis.* CUP.

IPCC. (2007b). *Climate change 2014: Impacts, adaptation, and vulnerability.* CUP.

IPCC. (2007c). *Climate change 2007: Mitigation of climate change.* CUP.

IPCC. (2013). *Climate change 2013: The physical science basis.* CUP.

IPCC. (2014a). *Climate change 2014: Impacts, adaptation, and vulnerability*. CUP.
IPCC. (2014b). *Climate change 2014: Mitigation of climate change*. CUP.
IPCC. (2019a). *Climate change and land*. In Press. www.ipcc.ch
IPCC. (2019b). *The Ocean and Cryosphere in a changing climate*. In Press. www.ipcc.ch
Jasanoff, S. (2004). *States of knowledge: The co-production of science and social order*. Routledge.
Jasanoff, S. (2010). Testing time for climate science. *Science*, *328*(5979), 695–696. https://doi.org/10.1126/science.1189420
Kuhn, T. (1962). *The structure of scientific revolutions* (3rd ed.). The University of Chicago Press.
Latulippe, N., & Klenk, N. (2020). Making room and moving over: Knowledge co-production, Indigenous knowledge sovereignty and the politics of global environmental change decision-making. *Current Opinion in Environmental Sustainability*, *42*, 7–14. https://doi.org/10.1016/j.cosust.2019.10.010
Lindroth, M., & Sinevaara-Niskanen, H. (2013). At the crossroads of autonomy and essentialism: Indigenous peoples in international environmental politics. *International Political Sociology*, *7*(3), 275–293. https://doi.org/10.1111/ips.12023
Long, S., Roberts, E., & Dehm, J. (2010). Climate justice inside and outside the UNFCCC: The example of REDD. *The Journal of Australian Political Economy*, *66*, 222–246. https://search.informit.org/doi/10.3316/ielapa.833621482268969
Longino, H. (2002). *The fate of knowledge*. Princeton University Press.
McAfee, K. (1999). Selling nature to save it? *Biodiversity and Green Developmentalism. Environment and Planning D: Society and Space*, *17*, 133–154. https://doi.org/10.1068/d170133
Miller, C. (2004). Climate science and the making of a global political order. In S. Jasanoff (Ed.), *States of knowledge: The coproduction of science and social order* (pp. 46–66). Routledge.
Miller, C., & Robbins, D. (2007). Cost-benefit analysis. In F. Fischer, G. Miller & M. Sidney (Eds.), *Handbook of public policy analysis: Theory, methods, and politics* (pp. 465–480). CRC Press.
Mortreux, C., & Barnett, J. (2009). Climate change, migration and adaptation in Funafuti, Tuvalu. *Global Environmental Change*, *19*(1), 105–112. https://doi.org/10.1016/j.gloenvcha.2008.09.006
Nadasdy, P. (1999). The politics of TEK: Power and the "integration" of knowledge. *Arctic Anthropology*, *36*(1/2), 1–18.
Nunn, P., & Reid, N. (2016). Aboriginal memories of inundation of the Australian coast dating from more than 7000 years ago. *Australian Geographer*, *47*(1), 11–47. https://doi.org/10.1080/00049182.2015.1077539
Nyong, A., Adesina, F., & Elasha, B. (2007). The value of indigenous knowledge in climate change mitigation and adaptation strategies in the African Sahel. *Mitigation and Adaptation Strategies for Global Change*, *12*(5), 787–797. https://doi.org/10.1007/s11027-007-9099-0
Petheram, L., Zander, K., Campbell, B., High, C., & Stacey, N. (2010). 'Strange changes': Indigenous perspectives of climate change and adaptation in NE Arnhem Land (Australia). *Global Environmental Change*, *20*(4), 681–692. https://doi.org/10.1016/j.gloenvcha.2010.05.002
Reichel-Dolmatoff, G. (1982). Cultural change and environmental awareness: A case study of the Sierra Nevada de Santa Marta, Colombia. *Mountain Research and Development*, *2*(3), 289–298. https://doi.org/10.2307/3673093
Said, E. (2003). *Orientalism*. Penguin.
Schroeder, R. (1999). Community, forestry and conditionality in the Gambia. *Africa*, *69*(1), 1–22. https://doi.org/10.2307/1161075
Semali, L., & Kincheloe, J. (1999). Introduction. In L. Semali & J. Kincheloe (Eds.), *What is Indigenous knowledge? Voices from the academy* (pp. 3–58). Falmer Press.
Shawoo, Z., & Thornton, T. F. (2019). The UN local communities and Indigenous peoples' platform: A traditional ecological knowledge-based evaluation. *Wiley Interdisciplinary Reviews: Climate Change*, *10*(3), 575. https://doi.org/10.1002/wcc.575
Smith, H. (2007). Disrupting the global discourse of climate change: The case of indigenous voices. In M. Pettenger (Ed.), *The social construction of climate change: Power, knowledge, norms, discourses* (pp. 197–216). Ashgate.
Solomon, M. (2006). Groupthink versus the wisdom of crowds: The social epistemology of deliberation and dissent. *The Southern Journal of Philosophy*, *44*(S1), 28–42. https://doi.org/10.1111/j.2041-6962.2006.tb00028.x

Suiseeya, K. (2017). Contesting justice in global forest governance: The promises and pitfalls of REDD+. *Conservation and Society*, *15*(2), 189–200. https://doi.org/10.4103/cs.cs_15_104

Suliman, S., Farbotko, C., Ransan-Cooper, H., Elizabeth McNamara, K., Thornton, F., McMichael, C., & Kitara, T. (2019). Indigenous (im)mobilities in the anthropocene. *Mobilities*, *14*(3), 298–318. https://doi.org/10.1080/17450101.2019.1601828

Taylor, M. (2015). *The political ecology of climate change adaptation*. Routledge.

Tschakert, P. (2007). Views from the vulnerable: Understanding climatic and other stressors in the Sahel. *Global Environmental Change*, *17*(3-4), 381–396. https://doi.org/10.1016/j.gloenvcha.2006.11.008

Turnhout, E., Dewulf, A., & Hulme, M. (2016). What does policy-relevant global environmental knowledge do? The cases of climate and biodiversity. *Current Opinion in Environmental Sustainability*, *18*, 65–72. https://doi.org/10.1016/j.cosust.2015.09.004

Turnhout, E., Neves, K., & de Lijster, E. (2014). 'Measurementality' in biodiversity governance: Knowledge, transparency, and the intergovernmental science-policy platform on biodiversity and ecosystem services (ipbes). *Environment and Planning A: Economy and Space*, *46*(3), 581–597. https://doi.org/10.1068/a4629

Van der Sluijs, J., Van Est, R., & Riphagen, M. (2010). Beyond consensus: Reflections from a democratic perspective on the interaction between climate politics and science. *Current Opinion in Environmental Sustainability*, *2*(5-6), 409–415. https://doi.org/10.1016/j.cosust.2010.10.003

Vasileiadou, E., Heimeriks, G., & Petersen, A. (2011). Exploring the impact of the IPCC assessment reports on science. *Environmental Science & Policy*, *14*(8), 1052–1061. https://doi.org/10.1016/j.envsci.2011.07.002

Verran, H. (2002). A postcolonial moment in science studies: Alternative firing regimes of environmental scientists and aboriginal landowners. *Social Studies of Science*, *32*(5-6), 729–762. https://doi.org/10.1177/030631270203200506

Watson, I. (2014). Re-centring first Nations knowledge and places in a terra nullius space. *AlterNative: An International Journal of Indigenous Peoples*, *10*(5), 508–520. https://doi.org/10.1177/117718011401000506

Weingart, P. (1999). Scientific expertise and political accountability: Paradoxes of science in politics. *Science and Public Policy*, *26*(3), 151–161. https://doi.org/10.3152/147154399781782437

Whyte, J. (2017). *Girls into science and technology: The story of a project* (Vol. 16). Taylor & Francis.

Wynne, B., & Lynch, M. (2015). Science and technology studies: Experts and expertise. In J. Wright (Ed.), *International encyclopedia of the social & behavioral sciences* (pp. 206–211). Elsevier Ltd.

Part II

Participation

Who wins and who loses from renewable energy transition? Large-scale solar, land, and livelihood in Karnataka, India

Devleena Ghosh, Gareth Bryant and Priya Pillai

ABSTRACT

This article evaluates the winners and losers of the energy transition playing out around the Pavagada solar park in Karnataka, India. We analyse how the innovative land leasing and competitive tendering mechanisms of the 'Pavagada model' have shaped the socio-ecological relations and outcomes of the project. We show the solar park has provided investors with low-risk returns, state electricity distributers with low-cost power, and larger landowners with new income streams. However, we find that the model is entrenching colonial, gender, and caste-based inequalities and oppressions, particularly for landless agricultural labourers who have lost their sources of livelihood. We argue that when renewable energy policy is primarily concerned with reducing costs and managing risks for private investors and state actors, the potential benefits of energy transition will not be equitably shared. The case of Pavagada demonstrates the need for democratic ownership and control of renewable energy.

Introduction

The transition away from fossil fuels towards renewable energy has unfolded unevenly across different communities, regions, and countries (Brown et al., 2015; Princen et al., 2015). Since the 2015 Paris Agreement, the Indian government has become an increasingly enthusiastic advocate for renewable energy, as the cost of renewable power has fallen well below that of coal and the sector has come to be seen as a new source of international investment. At the Glasgow climate conference, Prime Minister Modi announced a target of 500GW of non-fossil fuel energy by 2030, representing half of India's installed capacity.

With the rapidly falling costs for renewable energy, and with the possibility of emergent India-based production of renewable technologies, India's climate, energy, and industrial policies have converged on the sector. Renewable energy capacity, not including large hydropower, tripled since 2013 to 99GW installed capacity by July 2021, representing 26 per cent of total installed capacity. The rise of solar power has been particularly dramatic, rising almost 20-fold, from about 2.5GW in 2013 to 44GW in July 2021 (Central Electricity Authority, 2021; Kumar & Majid, 2020). Ambitious targets, even when they are not fully met, are encouraging the development of large-scale, centralized renewable energy, with wide socio-ecological implications.

Various policy models have emerged in renewable power financing and provision at the state level across India (Elavarasan et al., 2020). Examples include mass access to feed-in tariffs for farmers using solar water pumps in Gujarat, aggregated rooftop solar in Delhi, extensive use of small-scale wind power in Tamil Nadu, numerous examples of community-level power for micro-grids (Bhattacharyya et al., 2019), and the focus for this article, the use of land leasing arrangements and competitively tendered power purchase agreements for large-scale solar power in Karnataka. This state-level process of experimentation and development has advanced what has been characterized as India's 'energy transformation' (echoing the German 'Energiewende'; see Buckley, 2015) with pronounced regional heterogeneity.

Against this backdrop, anthropological, sociological, and geographical studies of the roll out of solar and wind projects in India and other contexts have demonstrated the need to critically examine the socio-ecological relations of renewable energy, particularly as they manifest in local communities surrounded by new energy infrastructures. This research has highlighted a range of issues concerning how the costs and benefits of renewable energy are distributed at the local level. These include conflicts over land use, variable employment opportunities and conditions, inequalities in electricity access, and new environmental pressures created along renewable energy supply chains (e.g. Hesketh, 2021; Stock, 2021; Yenneti et al., 2016; see Pearse & Bryant, 2021 for an overview).

In the Indian context, Cross's (2019a) study of decentralized solar technologies in rural Odisha reveals how electricity infrastructures – both on and off the grid – reflect and reproduce social inequalities such as those that exist between landowners and the landless, which are heavily shaped by caste and gender. As Cross (2019a, p. 78) argues, 'rural India's electric infrastructures do not just distribute "current"; they also distribute difference and inequality'. Comparable issues have been identified across technologies and geographies. For example, Boyer (2019), Howe (2019), and Hesketh (2021) have demonstrated how wind farm developments across Oaxaca, Mexico have generated social conflict and polarization over the capture of land rents by wealthy landowners, and the dispossession of Indigenous communities from their land.

A common thread in these critical perspectives is that the socio-ecological relations of solar and wind power are conditioned by the political economic context of energy provision and transition (Cherp et al., 2018; Christophers, 2022; McCarthy, 2015; Mulvaney, 2020; Newell, 2021; Webber et al., 2022). There are often tensions between utopian visions of renewable energy linked to promises of development and sustainability, and dominant models of renewable energy planning and financing (Boyer, 2019; Cross, 2019b). In the Indian context, Mookerjea (2019) highlights the 'multiple colonialisms' at play in an energy transition that remains closely tied to Eurocentric ideas about rationality and progress (Badami, 2021). Furthermore, the politics of renewable energy in India are inextricably linked to political claims for rights and recognition expressed through equitable and reliable access to electricity (Cross, 2019a). Thus, Coleman (2017, p. 33) encourages researchers to move beyond analysis of technical differences across time and space and to focus on the 'political terms' on which electricity is installed and operated in India.

This article builds on this literature by evaluating the winners and losers of energy transition, as it is playing out around the Pavagada solar park in the state of Karnataka, India. We draw on evidence gathered from field work conducted in Pavagada in 2018 and 2019, as well as analysis of policy and project documents. This research forms part of a larger international comparative project on the legitimacy of electricity decarbonization that combines ethnographic and political economic methods (Goodman, 2018).

The contribution of the paper is to demonstrate how the combination of mechanisms deployed in Pavagada – aggregating land through leasing and contracting electricity through competitive tendering – have shaped the socio-ecological relations and outcomes of the project. We argue that the case of Pavagada shows that when renewable energy policy is primarily concerned with reducing costs and managing risks for private investors and state actors, over a broader set of social, environmental, political, and economic goals, the potential benefits of energy transition will not be equitably shared. Discussions about just transitions commonly focus on justice for displaced fossil fuel workers. As the energy transition accelerates, justice in the production and distribution of renewable energy is becoming an equally pressing concern (Stephens, 2019; Sweeney, 2018). The case of Pavagada demonstrates that this requires democratic participation in and ownership over renewable energy planning and infrastructure.

The structure of the article is as follows. The first section outlines the evolution and status of solar power in India and Karnataka, including national- and state-level policy frameworks. The second section introduces the socio-ecological context of the Pavagada area and analyses the interaction between the innovative land leasing and competitive tendering mechanisms that constitute the 'Pavagada model'. The third section evaluates the socio-ecological relations of the model, focusing on the differentiated outcomes of the solar park for landowners, landless people, employment, women, and the local environment. The conclusion reflects on the need for renewable energy transitions to transcend fossil fuel modes of development.

Electrification and renewable energy in India and Karnataka

The recent development of solar energy in India has been shaped by the political–economic history of electrification in India. India's first prime minister, Jawaharlal Nehru, famously stated that 'electricity is perhaps the most necessary and the most revolutionary thing which you can take into the rural areas. The moment you take electricity, all kinds of things begin to move' (Kale, 2014, p. 1). Political visions of solar power articulated by recent Indian central governments have linked in with this tradition of regional development and national developmentalism. The Jawaharlal Nehru National Solar Mission, launched in 2010, laid down the norms for promoting solar energy as a major foundation of energy security, climate action, and access to electricity in India. As noted in the original Solar Mission policy, the decentralized possibilities of solar would 'ensure that people with no access, currently, to light and power, move directly to solar, leap-frogging the fossil fuel trajectory of growth' (SECI, 2010, p. 4). Prime Minister Narendra Modi leaned heavily on the promise of solar in his successful 2014 national election campaign as a basis for expanding electrification to rural areas. He pledged that every Indian household would have access to at least one indoor light powered by solar by 2019 (Katakey & Chakraborty, 2014).

The Solar Energy Corporation of India (SECI), the central government-owned enterprise tasked with implementing the National Solar Mission, has sought to achieve targets by promoting large-scale solar parks. It views large-scale solar parks as a way of concentrating transmission infrastructure and overcoming land bottlenecks that have emerged in more distributed generation models. To attract renewable energy developers to invest in large-scale solar parks, SECI has used competitive bidding mechanisms to auction power purchase agreements that lock in future sales of electricity. SECI has awarded power purchase agreements to 47 solar parks with a combined capacity of 25GW via its central auctions at rapidly declining prices (IBEF, 2021).

Electricity is regulated at both national and sub-national levels, and across India there is considerable variation in the energy mix between states. Kale (2014) showed that variations in levels of electrification between states, and the extent to which state governments have embraced market reforms, are due to different histories of political mobilization around power. The region that is now the state of Karnataka was one of the earliest in India to electrify. The government of the colonial era Princely State of Mysore, which largely overlaps with the territory of modern Karnataka, developed a system of state-owned hydropower in the early twentieth century as a strategy for developing its industrial base and raising state revenue (Kale, 2014, p. 37). This system of state ownership was carried on following independence through the framework for state-owned enterprises established by the Electricity Supply Act of 1948 (Kale, 2014, pp. 40–42). In addition to supporting industrial development, electrification was a state response to a strong rural movement demanding electricity for irrigation. The Karnataka government used a model of cross-subsidization between urban and rural energy users to electrify all irrigation pumps by the early 1990s – making it one of the first two states to achieve this, alongside Maharashtra (Kale, 2014, p. 90). Following successive market reforms in the 1990s and 2000s, the Karnataka government embraced the potential for private generation – first coal, then renewables – to support the growth of its IT hub city of Bengaluru (Kale, 2014, p. 127).

Karnataka is now the leading renewable energy state in India. 53 per cent of installed capacity on 31 July 2021 was renewable energy (not including hydropower) (Central Electricity Authority, 2021). Karnataka has enthusiastically pursued the development of renewable energy for four reasons. First, the state has favourable solar and wind conditions that create a large renewable energy potential. Second, the state was faced with a growing power supply deficit, with acute shortages at peak times. Third, the state aimed to attract private investment from international capital. Fourth, there was a need to reduce costs of supply to address financial stresses faced by embattled state-owned electricity supply companies (Buckley & Shah, 2018; Government of Karnataka, 2021).

To support renewable energy development, Karnataka has implemented various policy frameworks that have been focused on two main areas: first, securing the land needed for the installation of renewable energy projects and second, securing the market conditions needed for renewable energy developers to sell electricity. The state government has developed regulatory frameworks for land procurement via land acquisition or leasing. This has included changes to enable the automatic conversion of agricultural land for industrial purposes. The state government has developed policy mechanisms to underwrite revenue for generators selling power from renewable energy projects. These mechanisms include power purchase agreements that are either competitively tendered or negotiated with industrial energy users on a 'captive' basis, as well as renewable energy certificate schemes (Government of Karnataka, 2021).

With this suite of policies, Karnataka exceeded its 2021 renewable energy targets and looks set to exceed its 2030 target of 20GW of non-hydro-installed capacity, having already installed 15.6GW of capacity by July 2021 (not including large hydropower) (Government of Karnataka, 2021). This exceeds the state's combined installed capacity of coal (9.5GW), hydro (3.6GW), and nuclear (0.9GW) power and places Karnataka ahead of the other leading renewable energy states of Tamil Nadu (15.5GW), Gujarat (14.2GW), Rajasthan (11.4GW), and Maharashtra (10.4GW) (Central Electricity Authority, 2021; Ministry of New and Renewable Energy, 2021). Karnataka has also emerged as the leading state for solar power, with the highest installed solar capacity in the country, at 7.5GW (Ministry of New and Renewable Energy, 2021). A large portion of this is due to its flagship 2.05GW Pavagada solar park, which utilized land leasing and tendered purchase agreement policy mechanisms, and to which we now turn.

The Pavagada solar park 'model'

The Pavagada solar park is in the Pavagada taluk in the district of Tumkur, about four hours' drive north of the capital of Karnataka, Bengaluru. It is one of the largest solar projects in the world, occupying 12,000 acres of land. The solar farm encompasses five main villages (Balasamudra, Thirumani, Kyataganacharlu, Vallur, and Rayacharlu), centring on Thirumani, and supplies 2050 MW of installed capacity to India's national grid. It was established from 2015 through a collaboration between SECI and the state government-owned Karnataka Renewable Energy Development Ltd (KREDL). The two agencies established a joint venture company, the Karnataka Solar Power Development Corporation Ltd (KSPDCL), to manage the development of the project.

Socio-ecological context

Pavagada was chosen for the solar park because of its hot, semi-arid climate on an elevated plateau, its drought-affected agricultural economy, and the relative poverty of the population (Jain, 2015). Pavagada has minimal social and health infrastructure, poor schooling, a low literacy rate, and one of the lowest levels of college attendance in the state. Pavagada is therefore considered a 'more backward' taluk by the state government's 'development index' (Annigeri & Hanagodumath, 2018).

There were 10,294 people living in the five villages that surround the solar park at the time of the 2011 census. 2257 were cultivators (i.e. landowners) and 2580 agricultural labourers (i.e. landless workers) (Government of India, 2011). Traditionally, the landless populace works as agricultural labourers for landowners, and graze stock on private and common lands. Most landowners had small-to-medium holdings of less than 10 Ha, while there are a smaller number of larger landowners (Knight Frank, 2016, p. 38).

The stratification in land ownership intersects with caste and gender inequalities. Although most landowning farmers are men (1523 men and 734 women), the majority of landless agricultural labuorers are women (1409 women and 1171 men). Around 41 per cent of the population in the five villages are from Scheduled Caste or Scheduled Tribe communities (Government of India, 2011). Pavagada has been the site of conflict over land redistribution. This reflects local inequalities and relative disadvantage, and also its location near the border with Andhra Pradesh where there has been a long-running Maoist insurgency.

Agriculture in Pavagada has become increasingly precarious due to limited rain, leading to indebtedness, migration, and suicide among farmers. It has been officially declared drought-hit 54 times in the last 60 years, and the Central Water Board has declared its ground water is 'over-exploited' (Prabhu, 2017; Rajendran, 2019). Several officials we interviewed in Pavagada described the region to us as a 'virtual desert', or a 'wasteland' in making the case for the siting of the solar park. During our fieldwork during 2018 and 2019, we observed wide variation in agricultural conditions. Prior to our last visit in late 2019, there had been substantial rains, and the landscape looked relatively lush. We visited, for example, a well-established organic mango farm in a plot surrounded by solar panels. Farmers pointed out that this appearance belied general water scarcity and the rising impacts of climate change.

Land leasing

The Pavagada solar park's innovative land leasing model potentially represents a departure from other models of land acquisition used in India for industrial development. Government officials, cognizant of conflicts over 'land grabbing', proposed the land leasing model as a means of enrolling local farmers as long-term 'partners' in the project. In doing so, they were strategically responding to the distributed and uneven land ownership patterns in the area. Land aggregation (or 'banking', as it was described to us) was necessary because no single landowner could provide the quantity of land needed for a large-scale solar park. Leasing became the solution for this land aggregation problem because many landowners were unwilling to sell, due to long-held connections with the land. As a large landowner M who led negotiations explained to us:

> Whatever maybe the extreme condition, we are never willing to sell. It is kind of an attitude problem. It is an asset. If my father had sold, I wouldn't be having anything for me. It is your responsibility to maintain it.

At the opening ceremony for the park in March 2018, the Minister emphasized that the project treated farmers 'as the key partners, as also beneficiaries', and was 'creating new job opportunities and economic growth leading to the prosperity of the people of Pavagada' (Energy World, 2018).

The land leasing model was established as an entirely voluntary opt-in arrangement for local farmers who would retain ownership of their lands. Leases of land for the solar park were successfully negotiated within six months of landholders first being approached. Most landowners in the five villages leased their land to the project – 1422 in total. Of these, 96 larger landowners, with more than 10 Ha of land, provided 28 per cent of total land leased to the project. In contrast, 674 smaller landowners, with less than 2Ha of land, provided 15 per cent of total land leased to the project (Knight Frank, 2016, p. 38). The standard leasehold is a 28-year lease agreement that sets a base rate of Rs. 21,000 per acre per year (about USD300). This rate is indexed at a 'simple' interest rate of 5 per cent every two years, with indexation based on the original principal. As such, interest amounts to an additional payment of Rs. 1050 per acre every two years for the term of the lease. After 28 years, the lease can be renewed under mutual agreement between the landowner and KSPDCL. Otherwise, the land is to be returned in its original condition to the landowners, and there is a commitment to rehabilitate the lands for agricultural use, including to remove the solar panels and their concrete footings.

The leased land was levelled, and all vegetation removed before being divided into plots that could host 250MW of solar generation capacity each. The lease provides exclusive access to the land for KSPDCL and for the developers who sub-lease it. Developers enter a 25-year sub-leasing arrangement with KSPDCL, with periodic land rent, operational charges, and development fund contributions, creating a pool of funds to provide for infrastructure and local development needs (KSPDCL, 2020). KSPDCL, as broker for the land leases with both farmers and developers, has therefore been able to establish a new very large plant at very low public outlay.

Competitive bidding

Once access to land was secured via the land leasing model, plots were allocated to private investors through a competitive bidding process for power purchase agreements. This is an increasingly common policy model globally that seeks to both drive down the price of power and provide a government-backed income stream that developers can use to access capital (Baker, 2021; Bridge et al., 2020). 25-year power purchase agreements to supply electricity at a fixed rate per unit were 'reverse' auctioned to eight Indian and foreign large multinational companies by national and state agencies. The initial investors included Tata, Fortum, Adani, ReNew, Avaada, Softbank, and ACME, as well as KREDL, which maintains 50MW of capacity.

The price of power purchase agreements auctioned off has dramatically declined in a short period of time. The first set of auctions in 2017 resulted in contracts being awarded at 4.8 Rs/KWh. The cost of the power supplied to the grid in reverse auctions conducted in 2018 subsequently fell to 2.8Rs/

KWh. The contract for the final 50MW operated by the state-owned power company KREDL, negotiated in 2019, is even lower, at 1.2Rs/KWh (KSPDCL, n.d.; KREDL, 2019).

High investor interest despite rapidly declining prices shows the Pavagada model is an attractive one for private investors, who are anticipating a growing renewable energy market and falling technology costs. The model offers a long-term planning horizon with fixed costs and income, and low levels of forward pricing risk. The guaranteed power purchase price for 25 years is a major incentive in an international context where prices in power purchase agreements are often varied over time. Investors were also attracted to a new, innovative land leasing model that was promoted by KSPDCL as an opportunity to participate in what could become a more widely used model in the Indian solar economy.

The other main beneficiaries are the state-owned 'downstream' electricity supply companies (ESCOMs) that are allocated auctioned power purchase agreements. Karnataka has five ESCOMs, created through a series of market-oriented reforms that 'unbundled' various aspects of the electricity system, the largest of which is BESCOM, which supplies Bengaluru. Pavagada provides ESCOMs with access to lower cost power than was available from its previously coal-dominated energy mix. In the five years leading up to 2017, the average power purchase cost for Karnataka's ESCOMs ranged between 3.6 and 4.4Rs/KWh (Garg & Sundararagavan, 2018, p. 10). With the most recent Pavagada power purchase agreements heading below 2Rs/KWh, this represents a saving in power supply cost of around 50 per cent.

The cheap power supplied by Pavagada is significant due to the high levels of financial stress and indebtedness of the ESCOMs. ESCOMs have struggled to meet cost recovery goals imposed by national and state policy due to lack of metering infrastructure, particularly for irrigation pumps that represent 40 per cent of energy demand (Buckley & Shah, 2018; Garg & Sundararagavan, 2018). Lowering the purchase cost of power has made the subsidies provided to unmetered rural electricity users more financially sustainable for ESCOMs. This reflects a strategy identified by Kale (2014, p. 9) in states facing strong rural demands for electricity subsidies, which resist market reforms by 'finding alternative ways to augment electricity supply and utility profitability'.

Evaluating the Pavagada solar park

On the face of it, the Pavagada solar park contains the potential for significant social, economic, and environmental benefits for landholders, the Karnataka government, investors, distributors, and the climate. Reflecting this, the Pavagada model has been presented as an exemplar for other Indian states to follow. Does the Pavagada model live up to these high expectations? The next section of the paper examines the local community's experience of the solar park, drawing on our ethnographic evidence from interviews, workshops and meetings, and other relevant documents. We analyse the socio-ecological relations and outcomes of the project for landowners, landless labourers, employment, women, and the local environment, focusing on the views expressed by local people during our fieldwork.

The need to take proactive steps to ensure positive outcomes for the local community was recognized by social and environmental impact documents produced by the World Bank. One report clearly identified that the landless population were vulnerable to the loss of agricultural employment and access to livelihood:

> One of the critical issues would be the impact of the project on the agriculture labour/landless and especially on the marginalised sections of the society including the [Scheduled Caste] and [Scheduled Tribe] population (Knight Frank, 2016, p. 40).

With the solar park now fully operational, our research examined the distribution of outcomes among different social and economic groups – identified by the World Bank prior to the project development – in practice.

Landowners

Outcomes have differed for larger and smaller landowners. The minority of landowners with larger holdings have, in general, leased the bulk of their land to the solar park and retained some land, usually irrigated, for agriculture. Leasing land for the solar park offered an opportunity for larger landowners to diversify their income. Smaller landowners were more likely to lease all their land and could not pursue the same diversification strategy.

Previously, most farmers would earn between Rs.15,000 and Rs.25,000 per year per acre depending on the rains. Initially, the assured yearly lease rents of Rs. 21,000 per acre plus interest was generally seen as adequate. Landowners also agreed that there was no coercion to persuade them to enter the leasing arrangement. However, they emphasized that at the time of the negotiations, many were in serious financial stress. Negotiations were conducted after an extended period of drought and several farmers recounted that they needed a more reliable source of income. Landowner KA, who plays a prominent role on a committee of farmers who have leased their land, explained:

> While we were struggling due to no rains since 10 years, the political leaders came, held meeting and it became one sided decision for farmers. They told they give this much … They came, they asked, we were hungry, we gave it to them.

Although leasing the land provides landowners with a much more secure income than rain-contingent agriculture, many now held the view that the annual rent did not reflect an adequate compensation for the contribution their land was making to the project. This view emerged when landowners compared and calculated the economic returns from the solar park with the diminishing value of the leases. The simple, rather than compounding, rate of increase of the lease amount means that rental incomes may not keep pace with inflation over time. KA said people were questioning why farmers:

> gave away for such low price for a [multinational company]. They have made so much benefit … At present they are giving us [in lease rent] one day of their earning out of 365 days.

Landowners emphasized an adequate rent amount was needed because they were making a long-term commitment that reduced their alternative options as landholders. The re-zoning of land from agricultural to industrial use also has some potential financial disadvantages for landowners, who lost access to concessional agricultural loans. When asked what advice they would give to other communities being approached to lease their land for solar, landowner CR said:

> First, they should ask for increasing the rate. Then, while writing the agreement, they should show it to someone with education, and get it examined as to what has been included in the agreement. Because, I don't know how to read, I don't know English, I can only sign, that's all. If we have to understand the agreement, then we have to show it to someone like you.

This comment attests to the power imbalances that structured the negotiations, which left many landowners feeling as though they hadn't been given complete information.

Disconnection from the land generated some ontological fears among landowners. They thought that they might lose skills and knowledge over the 28-year lease term, making it difficult for the next generation to continue farming. Several farmers were very concerned that it was no longer possible to identify their lands in the levelled solar park. Landowner S3 reflected this sentiment:

> We feel our land was somewhere here, we are not able to recognise our own land ... I have worked in my land throughout my life; still I am struggling to keep my field memory alive. Then how will our children and family members remember. It is like delineating one's family from our roots, from our history is what bothers me the most.

Finally, the shift from farming to land leasing generated more immediate material change in reducing the capacity of farmers to grow food for their own needs. S2, a landowner who also works at the solar park, explained this shift, saying 'last year we had milk, buttermilk and curd at our house. Now we are buying'. KA described this in terms of the commodification of social relations, via increased market dependency:

> If we grew lentils for a year, we would use it for household consumption up to two years. Peas, millet etc, we used to grow and whatever is left after our share for household consumption, we used to sell it in the market. Now for every small thing, we are dependent of the market.

Landless labour

Despite its inadequacy, the lease income offers benefits by protecting the larger landholders, and to a lesser extent smaller landholders, from the uncertainties of agriculture. As anticipated by the World Bank report, local people who have no land, or have very little, are the main losers from the model. Describing the situation before the development of the solar park, one landless labourer (M4) said:

> I used to get work in the crop/agriculture season. You can say roughly five, six months of work. Only in peak summer, for four months there won't be agriculture work.

The landless population has historically relied on this seasonal agricultural labour, as well as very small-scale informal subsistence cultivation and access to common grazing lands. Referring to relatively open access given to landless labourers for grazing outside of peak growing seasons, P, a landless man said that this had been disrupted:

> Before solar it was very good for our sheep. After solar there are difficulties ... There are companies around us. There is no space for grazing.

The loss of local agricultural labour, and access to grazing land, has undermined the primary sources of livelihood for landless labourers, who are disproportionally Dalits and Adivasis (Scheduled Castes and Scheduled Tribes). One landless woman (V) described the loss of employment as a condition of 'idleness' and said they were surviving on small loans from larger landowners who had leased their land. Another, W1, said this was exacerbating inequality, because:

> People who had more land become wealthier. We are where we are, landless people.

This view was also widely held by landowners, who expressed to us their concern for the future for local landless people. A landholder with 3 acres of land (M2) said:

> People who do not have land, no benefit for them. Rather, they have lost their livelihood, as no one calls them for work. The landed people have given away their land for solar plant. A village means not just the landed agriculturist. There are all kinds of people, like people with livestocks, labourers, small farmers, people doing other jobs, etc. The government should support the villagers. But the situation is bad. The landless, poor people are forced to migrate and work outside.

Finally, one of the biggest frustrations among landless people was not only the loss of agricultural employment, but the perceived lack of infrastructural benefits from the solar park. Local villagers claimed to us that people 'like you' had come to the village and promised schools, hospitals, and water and power upgrades in addition to the solar park, but that these promises had not been kept.

Electricity was a particular cause of frustration due to the stark contrast between the solar and transmission infrastructure surrounding the villages, and the unequal and unreliable connections to the grid within the villages. The poorest households face problems of unpredictable load shedding and supply levels that are limited in many cases to being able to power a single lightbulb or appliance. One of our interviewees (M1) said:

> Initially when they came, they said they'll give free current (power) and filtered water. Those people told us before they even took the land … They also said they will build hospitals. Nothing is given. Lot of promises was made.

Employment

At initial meetings between officials and landowners about the land leasing model for the solar plant, the energy minister stated that the development would help address unemployment faced by farming communities in Pavagada. The solar park has generated some additional employment, but this has been unevenly distributed both among local people, and between local people and migrant workers from other parts of India. The project was projected to bring at least 1400 direct new solar jobs to the region plus indirect jobs in hotels, restaurants, and vehicle maintenance (PwC, 2015, p. 97). Local people stated there had been fewer jobs than predicted, and that many of these were only temporary, because they were limited to the construction phase, or intermittent, in the case of cleaning and grass-cutting work.

Limited ongoing job openings were generally allocated to landowners and their families. This reflects KSPDCL's Implementation and Support Agreement for the project, which states that 'first priority' would be given for family members of landowners who had leased their land (KSPDCL, 2018, p. 19). Landowning farmers said that access to these jobs was crucial for them to supplement their lease income. B, a landowning farmer with professional working experience, said:

> We need work. Otherwise we can't survive in the lease amount. Or else they have to increase the lease amount.

However, this policy excludes many of the landless workers, who are not benefiting from lease rents, from employment opportunities. The experience has been a source of disappointment and frustration for landless labourers, many of whom reported being promised employment opportunities. W1 said:

> Some people from solar park side, met us and promised that first preference will be given to landless people. Glass cleaning work will be given to you people … Energy minister promised work for everybody, especially landless people.

Workers, particularly security guards, which is one of the most consistent jobs available, report quite different levels of pay and working conditions between different companies operating in the solar park, which has resulted in regular labour disputes. Some companies are widely understood to be better or worse employers, based on their hiring practices, payment of wages, and access to other benefits such as health plans. This variation is in part due to the fact that working conditions and pay are not regulated via the main components of the Pavagada model: land leases and power supply contracts.

There is also general resentment from both landowners and landless labour for being excluded from higher paid, skilled jobs, by subcontracting arrangements that use internal migrant labour from states such as Orissa, West Bengal, and Jharkhand. B1 said:

> Now they bring, labourers from other states since these corporate project managers, have their own vendors who supply labour, resulting in commission for the managers. When youngsters ask for a job, they are often denied it. We are giving lands but not getting a job in return. We have skilled labour among us too, including diploma holders, [industrial training institute certificates], etc.

Training programmes that were included in the Implementation and Support Agreement as a way of overcoming local skill shortages had also not materialized (KSPDCL, 2018, p. 19). The key aspiration of local villagers to whom we spoke, especially local women, was for better jobs not just within, but most importantly beyond, the solar park.

Women

The loss of previous sources of income, whether agricultural income for landowners, or agricultural work for labourers, has disproportionally impacted women and their financial autonomy. For landless women in particular, the loss of open access grazing meant they had lost an important source of autonomous income. Similarly, many women who were small-scale landowners were also responsible for rearing sheep or goats, which provided a source of livelihood that they were able to exert some control over. N, a 37-year-old woman with 8 daughters, said that she had leased two acres of land to the solar park, but had been told that grazing would continue to be possible. She finds it increasingly difficult to maintain her flocks because the pastures and commons are now enclosed within the boundaries of the solar park. She was forced to sell some of her animals and migrate to nearby regions for 4 months of the year to graze her remaining animals.

Women we interviewed described how these changes entrenched gender inequalities. This issue was put to us forcefully in a meeting at a village where women conveyed their dissatisfaction with the loss in income. W1 said that since losing their sources of income, 'we have to depend on our husbands' income and they control us more'. Parallel issues emerged for women from landowning families. Women we interviewed protested that the process of negotiating land leases had placed male landowners in a privileged decision-making position. W1 said:

> They have not discussed with their families. The men went to meeting in Thirumani village. It was all men meeting! It was not a discussion over there, just agreeing to whatever they offered.

Women have also been less likely to find sufficient work to supplement rental income. A, a 31-year-old woman, said that the women in the village were excluded from the solar project as there were few jobs for women. She also mentioned that the male migrant labour who occupied the jobs available was viewed as a potential security issue for women, which enabled their families to further restrict their movement in an already patriarchal milieu. R, a 35-year-old woman who had leased 5 acres of land to the solar park, commented:

> Earlier women had agricultural work in the village and used to earn Rs 200 for sowing, weeding, harvesting and so on for 4–5 months a year and used to get paid once weekly on Wednesday. Every Wednesday there is a village market and they used to buy household stuff with the money and generally led an average life. Now, there is no work anywhere.

Women were particularly concerned about no longer being able to pay school fees and extremely worried about their children's future. One woman (W2) said:

> We do not want our children to lead a life like us. We want to give them better education, so that they can make a life for themselves ... Our aspiration is to educate our children and we cannot afford to send them to convent.

Local environment

Finally, there are several potential environmental concerns. There are concerns that meeting the park's water needs, including regular cleaning of the 400,000 panels, draws on drinking and irrigation water supplies at unsustainable levels. The Social Management Plan warned that 'the withdrawal of ground water through bore wells to meet the [operations and management] demand for proposed solar park would lead to adverse impacts on the already deteriorating ground water conditions' (Knight Frank, 2016, p. 54).

These impacts need to be compared with the water demands of pre-existing agricultural activities, both in terms of the quantity of water used, and the relative social and economic benefits of that water use. They also need to be compared with more sustainable ways of managing the water requirements of the solar park. Despite recommendations in these reports for water recycling and rainwater harvesting, the water consent agreement issued by the Karnataka State Pollution Control Board (2016) allows operators to withdraw water from bore wells without any requirement to mitigate the impact. KSPDCL representatives stated that some developers were using a water-less cleaning mechanism but since its use was not mandatory, and was relatively more complex to implement, not all companies had adopted it.

There are also concerns about the management of the local ecology surrounding the plant. Again, water is a key concern, because solar park construction had altered irrigation infrastructure. Farmer S2 said:

> The water level is too low, because to develop solar park, the company people have demolished the check dams. The rainwater flows away, they are not able to store the water and because of this the ground level water is also depleting.

These impacts have been unevenly distributed by caste. Dalit communities have been most heavily impacted because they are excluded from using water in the main village, and therefore depend on water supplied to common tanks (Saldanha & Rao, 2020).

All of the farmers to whom we spoke were concerned about the future health of their land and its condition after 28 years. They were worried about the impact of land clearing and levelling on future agricultural productivity, and were calling for money to be spent on afforestation. They were also concerned about the disposal of solar panels and other materials at the end of their life cycle and/or the lease term. Landowner S1, who leased 20 acres of land, viewed this environmental question through the lens of power imbalances:

> We are owners, but owners with no control over our land ... we have no control over the situation so will not be able to negotiate a better deal even after 28 years ... eventually the government and the companies will become powerful.

Conclusion

The Pavagada solar park has delivered much needed renewable energy to India's grid, which is critical for addressing the climate challenge. The innovative land lease model offers potential benefits for the development of solar power by addressing land bottlenecks while avoiding the need for controversial, often forced, land acquisitions. The current leasing and power purchase model is delivering substantial benefits for some stakeholders. Solar developers are guaranteed a return at low risk on their investments over a long-time horizon. State-owned power distributers are accessing a major supply of clean electricity at very low cost, potentially supporting the continuation of power subsidies for rural users. Large landowners are able to diversify their sources of income with rental payments underpinned by the state government.

Yet, the Pavagada model is unequally distributing the benefits of renewable energy, while entrenching existing social inequalities. The majority landless population have experienced a loss of income from agricultural employment on land that has been leased, which has not been fully replaced by access to new jobs in the solar park or elsewhere. Women have suffered the greatest losses because they are over-represented as landless workers and have lost income from agricultural work that was a source of financial independence. Both landowners and landless have experienced a loss of connection to the land that was a key part of their collective identity, posing questions about future livelihood and vocation, both for themselves and for future generations. Broader social and economic development opportunities that could be underpinned by the solar park are yet to materialize.

The Pavagada model, combining land lease aggregation and competitive bidding mechanisms for power purchase agreements, is reflective of a globally dominant mode of energy transition that is primarily focused on driving down costs and managing risks for investors and states. There are, however, some relatively simple changes to the Pavagada model that could improve social, economic, and environmental outcomes. These include ensuring that land leases are properly valued and indexed, and that power purchase agreements require local employment of landless people, prioritizing women, and those from lower caste and Adivasi backgrounds. From the perspective of private investors and state actors, these represent costs and risks that need to be absorbed. However, this is necessary if Pavagada and other large-scale renewable energy projects that adopt a similar model are to generate real and shared benefits for local communities.

The difficulties faced by people in Pavagada are, of course, structured by a much longer history of exploitation and colonization that is common across rural India. This is recognized by local community members. KA explained:

> We are not saying that we have 100% difficulties because of [the solar park]. Earlier the crop yield was very less and now we somehow lead our life. But we do not have better life.

However, the Pavagada solar park represents a missed opportunity for transition that would leverage renewable energy development to create better lives for disadvantaged communities.

Over the course of our research, it was clear that communities had thought deeply about what would be required to achieve more transformative change. The most common idea proposed was to move towards a more substantively democratic partnership model. Key components of this model would be a revenue sharing agreement, and greater scope for the co-management of land, to allow greater co-existence of electricity production and agriculture, and encourage better employment outcomes. According to KA, this was informed by a sense of permanency of Pavagada's role in the energy transition:

> The energy production will continue, the structures cannot be dismantled. Therefore, they should give us partnership.

The Pavagada solar park has transformed the socio-ecological relations of energy generated in Karnataka by decarbonizing its electricity supply. However, it has not delivered the wider transformation demanded by local communities that could represent a clear break from fossil fuel modes of development by challenging interlocking colonial, gender, and caste-based inequalities and oppressions. As Boyer (with Howe) (2019, p. 194) has argued 'it is all too easy for renewable energy development to occur with little or no social, political, or economic transition attached to it'. Yet, as with fossil fuel-based energy production, these processes are always contested (Ghosh, 2016), and activists on the ground are acutely aware of what is at stake. As R, an Indian activist for climate democracy we interviewed, put it:

> These stories are not any different from fossil-fuel stories, when we go to coal-bearing states or energy hubs of India. Land-issues, water-issues, livelihood-issues, job-issues etc are present there too. How are we going to come up with the vision statement, looking it ... not as a transition but make it into a transformation?

Acknowledgements

Many thanks to Kavitha Srinivasan for translations, and to the rest of the Decarbonising Electricity team for the collective research effort.

Disclosure statement

No potential conflict of interest was reported by the author(s).

Funding

This work was supported by the Australian Research Council under the Australian Research Council Discovery Project DP180101368: Decarbonising Electricity: A Comparison in Socio-ecological Relations.

References

Annigeri, V., & Hanagodumath, S. (2018). *Regional imbalances in Karnataka revisited: Reflections on financing for a balanced regional development*. Centre for Multi-Disciplinary Development Research (CMDR). http://cmdr.ac.in/editor_v51/assets/Mono-74.pdf

Badami, N. (2021). Let there be light (or, in defense of darkness). *South Atlantic Quarterly*, *120*(1), 51–61. https://doi.org/10.1215/00382876-8795706.

Baker, L. (2021). Procurement, finance and the energy transition: Between global processes and territorial realities. *Environment and Planning E: Nature and Space*. Online early. https://doi.org/10.1177/2514848621991121.

Bhattacharyya, S. C., Palit, D., Sarangi, G. K., Srivastava, V., & Sharma, P. (2019). Solar PV mini-grids versus large-scale embedded PV generation: A case study of Uttar Pradesh (India). *Energy Policy*, *128*, 36–44. https://doi.org/10.1016/j.enpol.2018.12.040

Boyer, D. (2019). *Energopolitics: Wind and power in the Anthropocene*. Duke University Press.

Bridge, G., Bulkeley, H., Langley, P., & Van Veelan, B. (2020). Pluralizing and problematizing carbon finance. *Progress in Human Geography*, *44*(4), 724–742. https://doi.org/10.1177/0309132519856260

Brown, L., Larsen, J., Roney, J. M., & Adams, E. E. (2015). *The great transition: Shifting from fossil fuels to solar and wind energy*. Earth Policy Institute, W.W. Norton.

Buckley, T. (2015). *India's energy sector transformation*. Institute for Energy Economics and Financial Analysis.

Buckley, T., & Shah, K. (2018). *Karnataka's renewable energy transformation*, Sydney IEEFA.

Central Electricity Authority. (2021). *National Power Portal*. Ministry of Power, Government of India, https://npp.gov.in/publishedReports, updated July 2021

Cherp, A., Vinichenko, V., Jewell, J., Brutschin, E., & Sovacool, B. (2018). Integrating techno-economic, socio-technical and political perspectives on national energy transitions: A meta-theoretical framework. *Energy Research & Social Science*, *37*, 175–190. https://doi.org/10.1016/j.erss.2017.09.015

Christophers, B. (2022). Fossilised capital: Price and profit in the energy transition. *New Political Economy*, *27* (1), 146–159. https://doi.org/10.1080/13563467.2021.1926957

Coleman, L. (2017). *A moral technology: Electrification as political ritual in New Delhi*. Cornell University Press.

Cross, J. (2019a). The solar good: Energy ethics in poor *markets*. *Journal of the Royal Anthropological Institute*, *25*(S1), 47–66. https://doi.org/10.1111/1467-9655.13014

Cross, J. (2019b). No current: Electricity and disconnection in rural India. In S. Abram, B. R. Winthereik, & T. Yarrow (Eds.), *Electrifying anthropology: Exploring electrical practices and infrastructures* (pp. 66–81). Bloomsbury Academic.

Elavarasan, R., Shafiullah, G. M., Padmanaban, S., Kumar, N. M., Annam, A., Vetrichelvan, A. M., Mihet-Popa, L., & Holm-Nielsen, J. B. (2020). A comprehensive review on renewable energy development, challenges, and policies of leading Indian states with an international perspective. *IEEE Access*, (4 May), 74432–74457. https://doi.org/10.1109/ACCESS.2020.2988011. https://ieeexplore.ieee.org/stamp/stamp.jsp?arnumber = 9072152.

Energy World. (2018, March 2). World's largest solar park launched in Karnataka. *Energy World*.

Garg, R., & Sundararagavan, S. (2018). *Karnataka distribution sector landscape*. Shakti Sustainable Energy Foundation. https://shaktifoundation.in/wp-content/uploads/2019/02/Karnataka-Distribution-Sector-Landscape-2.pdf

Ghosh, D. (2016). 'We don't want to eat coal': Development and its discontents in a Chhattisgarh district in India. *Energy Policy*, *99*, 252–260. https://doi.org/10.1016/j.enpol.2016.05.046

Goodman, J. (2018). Researching climate crisis and energy transitions: Some issues for ethnography. *Energy Research & Social Science*, *45*, 340–347. https://doi.org/10.1016/j.erss.2018.07.032

Government of India. (2011). 2011 *Census*, Population Data Tumkur District, Table DDW PCA 2917-2011, Household Data Tumkur District, Table HLPCA-29571-2011.

Government of Karnataka. (2021, October 13). *Draft Karnataka Renewable Energy Policy 2021-2026*. https://kredl.karnataka.gov.in/storage/pdf-files/Scrollfiles/Draft%20Karnataka%20Renewable%20Energy%20Policy%202021-2026_13%20Oct%202021.pdf

Hesketh, C. (2021). Clean development or the development of dispossession? The political economy of wind parks in southern Mexico. *Environment and Planning E: Nature and Space*. Advance online publication. https://doi.org/10.1177/2514848621991764

Howe, C. (2019). *Ecologics: Wind and power in the Anthropocene*. Duke University Press.

IBEF. (2021, June). *India Renewable Energy Industry Report*. India Brand Equity Foundation (IBEF). https://www.ibef.org/industry/renewable-energy.aspx.

Jain, A. (2015, December 22). Solar energy to bring jobs and prosperity back to parched villages. *World Bank Blog*. http://blogs.worldbank.org/energy/solar-energy-bring-jobs-and-prosperity-back-parched-villages-india

Kale, S. (2014). *Electrifying India: Regional political economies of development*. Stanford University Press.

Karnataka Renewable Energy Development Limited (KREDL). (2019). *Details of short term tender notification for the selection of EPC contractor for the implementation of 50MWAC* (2*10MWAC+2*15MWAC) Capacity Solar Power Project at Pavagada Solar Park. https://kredlinfo.in/Solargrid_50MW

Karnataka Solar Power Development Corporation Limited (KSPDCL). (2018, April). *Implementation and support agreement for grid connected solar photo voltaic projects for Block No. – 2000 MW Pavagada Ultra Mega Solar Park in Karnataka*. http://www.kspdcl.in/sp_docs/FINAL%20ISA.pdf

Karnataka Solar Power Development Corporation Limited (KSPDCL). (2020). *Charges payable to KSPDCL by solar power developers selected by KREDL (1200MW) and SECI (200MW)*. http://kspdcl.in/sp_docs/Charges%20payable%20by%20SPDs%20to%20KSPDCL%20for%20each%20block%20of%2050MW.pdf

Karnataka Solar Power Development Corporation Limited (KSPDCL). (n.d.). *Allocation to ESCOMs from Pavagada Solar Park.* http://www.kspdcl.in/SP_DOCS/Allocation%20to%20ESCOMs%20from%20pavagada%20solar%20park.xlsx

Karnataka State Pollution Control Board. (2016). *Consent order no: CTE-102527.* http://www.kspdcl.in/SP_DOCS/NOC%20from%20KSPCB%20dated%2031-05-2016.pdf

Katakey, R., & Chakraborty, D. (2014, May 19). Modi to use solar to bring power to every home by 2019. *Bloomberg.* https://www.bloomberg.com/news/articles/2014-05-19/modi-to-use-solar-to-bring-power-to-every-home-by-2019

Knight Frank (2016). *Social Management Plan– Solar PV Park Nagalmadike Hobli, Pavagada Taluk, Tumkur,* http://www.kspdcl.in/SP_DOCS/SMP%20Solar%20UMPP%20Pavagada.pdf

Kumar, C., & Majid, M. (2020). Renewable energy for sustainable development in India: Current status, future prospects, challenges, employment, and investment opportunities. *Energy, Sustainability and Society, 10*(2). https://doi.org/10.1007/978-981-15-0410-5

McCarthy, J. (2015). A socioecological fix to capitalist crisis and climate change? The possibilities and limits of renewable energy. *Environment and Planning A: Economy and Space, 47*(12), 2485–2502. https://doi.org/10.1177/0308518X15602491

Ministry of New and Renewable Energy. (2021). *Programme/Scheme wise Physical Progress in 2020-21 & Cumulative up to July.* https://mnre.gov.in/the-ministry/physical-progress

Mookerjea, S. (2019). Renewable energy transition under multiple colonialisms: Passive revolution, fascism redux and utopian praxes. *Cultural Studies, 33*(3), 570–593. https://doi.org/10.1080/09502386.2019.1585464

Mulvaney, D. (2020). *Sustainable energy transitions: Socio-ecological dimensions of decarbonization.* Palgrave Macmillan.

Newell, P. (2021). *Power shift: The global political economy of energy transitions.* Cambridge University Press.

Pearse, R., & Bryant, G. (2021). Labour in transition: A value-theoretical approach to renewable energy labour. *Environment and Planning E: Nature and Space.* Advance online publication. https://doi.org/10.1177/25148486211055542

Prabhu, N. (2017, September 30). 1st phase of Pavagada solar park to be ready by Dec. *The Hindu.* https://www.thehindu.com/news/national/karnataka/1st-phase-of-pavagada-solar-park-to-be-ready-by-dec/article19779170.ece

Princen, T., Manno, J. P., & Martin, P. L. (2015). *Ending the fossil fuel era.* The MIT Press.

PwC. (2015). *Solar Park of 2000 MW capacity in the State of Karnataka, India.* Detailed Project Report: DPR. https://cdkn.org/wp-content/uploads/2017/07/Up-Karnataka-Solar-park-DPR-1.pdf

Rajendran, S. (2019). Emergence of the largest solar power plant in Karnataka – A microscopic analysis. *Journal of Environmental Nanotechnology, Monitoring and Management, 8*(1), 47–51. https://doi.org/10.13074/jent.2019.03.191351

Saldanha, L. F., & Rao, B. S. (2020). *Notes for the India Workshop of Governance of SocioTechnical Transformation Project (GoST) ESG Notes for GoST Workshop* – January 2020. https://esgindia.org/new/research/governance-of-socio-technical-transformations/notes-for-the-india-workshop-of-governance-of-socio-technical-transformation-project-gost/

Solar Energy Corporation of India (SECI). (2010). *Jawaharlal Nehru National Solar Mission: Towards Building Solar India.*

Stephens, J. C. (2019). Energy democracy: Redistributing power to the people through renewable transformation. *Environment: Science and Policy for Sustainable Development, 61*(2), 4–13. https://doi.org/10.1080/00139157.2019.1564212

Stock, R. (2021). Praeclariat: Theorising precarious labour geographies of solar energy. *Antipode, 53*(3), 928–949. https://doi.org/10.1111/anti.12698

Sweeney, S. (2018). *Another energy is possible,* Volume 44.2 of the Publication Series Ecology, edited by Heinrich Boll Foundation.

Webber, S., Nelson, S., Millington, N., Bryant, G. & Bigger, P. (2022). Financing reparative climate infrastructures: Capital switching, repair, and decommodification. *Antipode.* Advance online publication. https://doi.org/10.1111/anti.12806

Yenneti, K., Day, R., & Golubchikov, O. (2016). Spatial justice and the land politics of renewables: Dispossessing vulnerable communities through solar energy mega-projects. *Geoforum; Journal of Physical, Human, and Regional Geosciences, 76,* 90–99. https://doi.org/10.1016/j.geoforum.2016.09.004

Legitimizing energy transitions through community participation: Germany and Australia at a crossroad

Franziska Mey, Anne Kallies, John Wiseman and Mitchel Watson

ABSTRACT

Public acceptance is central to the success of a transition agenda but with the increasing uptake of renewable energy technology, achieving and sustaining the public legitimacy for accelerating the energy transition is at risk. This study provides an updated account of the energy transition development in Germany and Australia with a focus on institutional legitimation. Drawing on the distinction between input and output legitimacy, we query how enabling community participation can enhance the legitimacy of policy instruments. We employ two case studies of energy transitions – those of Germany and Australia – to explain where and how community participation has supported the legitimacy of policy interventions. While input legitimacy remains high in both Germany and Australia, both examples illustrate how policy instruments can lack legitimacy if decision-makers lose sight of or never properly consider the output dimension of legitimacy.

1. Introduction

Energy transitions in their various forms have become central to addressing accelerating climate change. Worldwide, almost every country has established policy measures to foster the use of renewable energy generation (REN21, 2021). Acceptance of these measures are central to the success of a transition agenda. However, with the increased rollout of renewables and their socio-technical impact, public opposition is growing – and this is threatening the sustained legitimacy of the energy transition at large.

Indeed, a lack of legitimacy can jeopardize the global transformation of the energy system. Hence better understanding of different country pathways and their contributions and failures to generate public support for specific policy measures provides important insights for legitimizing energy transitions. We here employ two country case studies – that of Australia and Germany – to illustrate how governance frameworks can support or hinder the legitimation of an energy transition. The case studies illustrate the different transition pathways and policy fora in which legitimacy for the energy transition is created, sustained, or eroded.

Australia and Germany are two countries in the process of decarbonizing their energy systems at quite different pace and timing. The German energy transition goes back to the 1970s and is strongly linked to the anti-nuclear and alternative movements initiating the first collective renewable energy projects (Mautz et al., 2008). Australia followed some 20 years behind in its renewable energy development driven mainly by top-down policies, implemented through a nominally competitive market-oriented approach (Buckman & Diesendorf, 2010). At the time of writing, both countries appear to be at a crossroad. Germany has doubled its renewable electricity generation to 46% in the last 10 years (BMWi, 2021), Australia produced 27.7% from renewables in 2020 (Clean Energy Council, 2021) but is rapidly accelerating its renewable fleet. In both countries, the energy systems transformation is supported by a public majority. A 2021 report on Australia's attitude towards climate change and energy shows that 82% of Australians support the phase-out of coal and are concerned about climate change impacts (The Australia Institute, 2021). In Germany, a 2021 survey revealed that 83% of Germans support the further expansion of renewable energy (AEE, 2022). While this is a high number, the support by Germans for the expansion of renewables has steadily dropped from a high of 93% in 2017 (AEE, 2015). In the same survey, questions on whether renewable energy installations were acceptable in the immediate neighbourhood, acceptance dropped from 66% in 2015 (AEE, 2015) to 55% in 2021 (AEE, 2022). Specific policy interventions and associated infrastructure projects, such as grid expansion and the locating of wind farms are contested in both jurisdictions.

The comparison highlights participation in different forms and through a variety of governance frameworks. How and where in legal framework participation gets supported is often historically grown, and the authors have therefore chosen a case study rather than a side-by-side comparison approach to highlight the unique nature of each of the transitions underway in the example jurisdictions of Germany and Australia.

Drawing on the distinction between input- and output-legitimacy, and in particular, the role of community participation in creating and sustaining output legitimacy, our key research questions are:

How is community participation enabled in the institutional frameworks of our case studies of Germany and Australia and what was the role of community participation for legitimizing energy transition policies in the past?

Clearly, renewable energy transitions can create winners and losers in the community, with potentially severe impacts on the legitimacy, justice, and fairness of the energy transition. The ongoing Covid-crisis, the impact of the war in Ukraine on European but also global energy markets, as well as the now keenly felt impacts of climate change on electricity systems, all impact on energy transitions. A continuous reflection and discussion of these issues remain pertinent.

This study provides an updated account of the energy transition development in Germany and Australia with a focus on institutional legitimation, which shows how developments in the past are leading to current trends and new challenges. We address the different ways in which legitimacy has been achieved in the respective energy transitions. We assess how different policy developments provide for participatory processes and show how legitimacy can be enhanced, or lost, in policy changes. We argue that legitimacy outcomes are significantly impacted by the role 'carved out' for community participation in the implementation of transition policies.

The paper is structured as follows: In Part II, we introduce the concept of legitimacy in the context of energy transitions. Parts III and IV assess the developments in Germany and Australia, respectively. Part V summarizes and concludes.

2. The concept of legitimacy in the energy transition

The legitimacy of the energy transition is a widely discussed and recognized concept. The term often refers to public acceptance of transition agendas and measures or specific technologies, in particular renewable technologies, associated with the energy transition agenda. The just transition discourse also mirrors the importance of public support and legitimizing policy measures. We here provide an overview of the different conceptualizations of legitimacy provided in the energy transition literature and explain the concept of input and output legitimacy (first coined by Scharpf, 1999), and in particular, the role of community participation in enhancing the latter, which will inform the structure of the rest of the paper.

One stream of energy transition literature highlights the importance of legitimacy in the context of technology adoption (Markard et al., 2016). They argue that legitimacy constitutes a prerequisite to mobilize financial, human and material resources as well as regulatory support and is particularly vital for novel technologies, especially if they are radically different from existing ones (Aldrich & Fiol, 2007; Bergek et al., 2008; Harris-Lovett et al., 2015). The authors claim that legitimacy is neither given nor purely emergent (Markard et al., 2016). Instead, it is created in a collective, social process that remains subject to contingencies and conflict with context structures changing over time. Sareen and Haarstad (2020) conclude that legitimacy is a deeply conflicted issue where various actors negotiate and struggle for control. Fuchs (2019) shows how actors legitimate novel organizational forms in the emerging renewable energy sector in Germany. He considers legitimation as a relational category referring to the breakup of existing relations and an abandonment of accompanying frames.

In this paper, we draw on institutional legitimation – in particular, the role of governance frameworks in legitimizing transitions. A common thread running through different descriptions of the term 'legitimacy' is 'acquiring public support of a measure'. An early account by Weber (1978) describes legitimacy as 'the acceptance of exercised power', in other words, a general willingness to follow the commands of a ruler or ruling body. Political science distinguishes between two dimensions of legitimacy: input and output (Scharpf, 1999). Scharpf (1999) states that the input-oriented perspective considers a political decision as legitimate if it reflects the will of the public. The output-oriented perspective, on the other hand, considers policy decisions as legitimate if they are made 'for the people' and represent 'effective solutions to common problems of the governed' (Scharpf, 2003).

Applying the concept to particular policy measures seeking to support can be ambiguous: is it the technology use as such, is it the spatial distribution, i.e. the locations of installations (distributive dimension) or the associated planning, procedures and processes (processual dimension), which needs to be legitimized? Other ways of looking at this comprise the perspectives of communities and their circumstances (recognition dimension), individual or collective engagement (input) and the impact on the result (outcome) (Radtke, 2020). Is it important that the public actively supports energy transition measures, or is the absence of opposition sufficient to claim legitimacy?

Vringer and Carabain (2020), in their assessment of the legitimacy of Dutch transition policy, refine Scharpf's concept of input and output legitimacy for the energy transition. In their conception, input legitimacy describes *public support for policy goals*, while output legitimacy is *public support for the intervention seeking to achieve the policy goal* – a definition adopted in this paper. To apply this definition to energy transition efforts: Input legitimacy describes the societal consent for the transition as such. As shown in the introduction, input legitimacy remains high in both Germany and Australia. Vringer and Carabain (2020) emphasize that public support for climate action is not sufficient to achieve legitimacy but achieving policy goals will also require support for specific policy interventions. Output legitimacy attaches to the legitimacy of employing particular policy instruments to achieve this transition outcome – be it renewable energy support schemes, Feed-in-Tariffs or targeted intervention through planning frameworks.

How then can output legitimacy be assessed? For the purpose of this paper, we draw on a central theme in energy transitions literature – that of community participation and its role in providing legitimacy to interventions. Particularly since participatory approaches to the energy transition can help to build a long-term legitimation of the transition. Participation is understood as all forms of public involvement along the so-called 'Ladder of Citizen Participation' (Arnstein, 1969), on which several levels of intensity of participation are mapped: from non-participation to complete decision-making power in the hands of citizens at the top level. In the context of energy projects, participation is associated both with the process (actors' involvement from concept to implementation and operation) as well as the outcome (e.g. financial benefits) dimension (Walker and Devine-Wright, 2008; Creamer et al., 2019). Community-owned renewable energy (in forms of different business models such as cooperatives, not-for-profit organizations and social enterprises) is often considered as a proxy for higher levels of participation, where local actors are involved or drive a local energy project and benefit from it which ultimately increases the legitimacy of the initiative (Cowell Bristow, & Munday, 2011; Gross, 2007; Hicks & Ison, 2018; Jenkins et al., 2018).

Wüstenhagen et al. (2007) describe the related concept of social acceptance as crucial to energy transitions success – distinguishing three dimensions – socio-political, community and market acceptance. They emphasize the importance of how and when communities get involved, impacts of local economy and amenity values, as well as issues of perceived justice, fairness as well as trust in the decision-making process. Citizen, or community energy research has shown that community involvement can enhance social acceptance or *legitimacy* of policy measures seeking to achieve transition outcomes.

The related energy justice literature can inform our discussion of output legitimacy. Energy justice research has identified that fairness of benefits, costs, and the procedures for how the distribution of such is achieved, are critical for the legitimacy of energy transition outcomes (Heffron & McCauley, 2017; Jenkins et al., 2017; Jenkins et al., 2018; Mundaca et al., 2018). Studies have revealed deficits in legitimacy particular in regard to acceptance of and participation in the energy transition (Hildebrand et al., 2018).

However, there are differences in country approaçhes to address and achieve output legitimacy. Their governance frameworks can emphasize top-down or bottom-up measures, balancing tensions between centralized features, such as managing a shared grid, and global drivers, like the Paris Agreement, with the actual implementation of energy policy and its local impact. As MacArthur (2016) shows, legitimate participation is not just a question of which actors are involved in decision-making, but is also determined by instrument choice. MacArthur states that instrument choice is '*far from a simple technical matter*'; but can reveal commitments to citizen empowerment or, indeed, a lack thereof. We, therefore, question the opportunities policy choices in Australia and Germany provide for meaningful participation.

As a final comment, a transition *towards* a high renewable future is, at the same time, a transition *away* from a different legacy fossil-fuel based energy system. While we in this paper focus on the former, it should be acknowledged that for both Germany and Australia, the successful exit from the use of coal (and nuclear power) for electricity generation will be central for their respective transitions. Legitimacy of a transition, therefore, will not only depend on how well renewable energy, and its associated infrastructure and changes to energy use patterns, are received by the public, but also, to what degree the coal-phase is designed in a just and participatory way. As Wiseman and Wollersheim (2021) show, multi-stakeholder engagement and meaningful participation will provide legitimacy for a managed transition away from coal. A rich literature is addressing coal exit policies and can provide further insights (Coenen et al., 2018; Harrahill & Douglas, 2019; Snell, 2018).

3. Energy transition in Germany – community participation as driver

The German Energiewende is often held up as the model for a successful transition of the energy system. Indeed, the German transformation project characterized by a combination of bottom-up and top-down measures has received worldwide attention for its seemingly smooth pathway to achieving a rapid and well-managed energy transition (Holstenkamp & Radtke, 2018). It is important to note that community energy was at the heart of early transition efforts (Mautz et al., 2008; Yildiz et al., 2015).

Germany embarked on changing its electricity system more than 30 years ago. However, it was only in the last decade that the increasing renewable energy generation (from 6% in 2000, 20% in 2011 to 46% in 2020) changed the power dynamic in the national electricity system, which ultimately supported the discussion about phasing-out nuclear and coal-fired power stations. Indeed, many studies have highlighted the key success factor facilitating the German transition process being its actor diversity and bottom-up approach (Mautz et al., 2008; Mey, 2017; Yildiz et al., 2015). The energy transition has, however, arguably entered a difficult period. With the reduction of the Feed-in-Tariff, solar PV lost its economic attractiveness, and the much-praised citizen energy movement largely came to a halt. Opposition emerged from a new anti-wind movement, (new) political parties, and nature conservation organizations while the much-needed grid extension and increasing electricity prices exacerbated the conflicts (Nanz & Kamlage, 2014; Radtke, 2020; Schweizer-Ries et al., 2016).

Of course, with close to 50% renewable energy the transition is far from complete, and the last years have seen the transition process faltering. Hence crucial questions emerge about what legitimized the Energiewende in the first place? What diminished legitimacy recently and what mistakes are to be avoided?

3.1. Bottom-up drivers of change – a transition rooted in participation

The German energy transition has a long history and is strongly associated with different forms of individual and collective participation in energy deployment. Indeed, external events (Chernobyl, and later Fukushima) and pressure from the grassroots level created the impulse as well as the legitimacy for political action leading to the rise of renewable energy and the 'Energiewende' in Germany almost 30 years ago (Mautz et al., 2008; Morris & Jungjohann, 2016).

Community energy actors emerged from the anti-nuclear and environmental movements in the 1970s with a new focus and solution-orientated emphasis to explore alternative energy opportunities (Mey, 2017). The collective approach to deploy new innovative energy sources from the grassroots level was distributed broadly when policy interventions such as the Feed-in-Law (1990) and the Feed-in-Tariff (Renewable Energy Act, 2000) were introduced. These laws and regulations removed market entry barriers, enabled grid access and ultimately incentivized the development of solar, bioenergy and wind projects for small and medium actors such as homeowners, small businesses and collective actors. Indeed, Kahla et al. (2017) showed that between 1995 and 2016 more than 1700 collective community ownership structures emerged involving thousands of citizens in some form of energy deployment. The sector reached a peak between 2011 and 2014 (Kahla et al. 2017), demonstrated, for example, in the drop of new energy cooperatives since 2012 (Figure 1).

Another measure supporting local participation was the regulatory requirement (Building Code and Spatial Planning Act) to formally involve the local community (particularly in rural areas) in

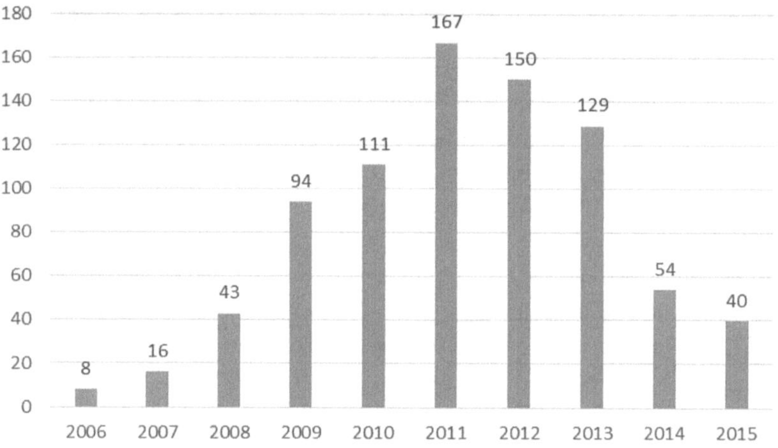

Figure 1. Growth of energy cooperatives in Germany – new establishments per year. Source: DGRV (2016).

the land selection process for the siting of large-scale infrastructure projects, e.g. wind turbines. This generally comprises information and consultation procedures facilitated by local authorities (Schneider & Boenigk, 2012). While this formal procedure did not contribute to greater community engagement as such (Wunderlich, 2012), it can trigger informal participation measures. While these depend on the good will of the responsible authorities (Kamlage et al., 2014), examples included local governments who recognized the importance of local acceptance of these new developments and started to actively engage their local residents in the development and implementation of local renewable energy strategies (AEE, 2012, 2013). This often also involved the initiation and facilitation of community-owned renewable energy projects.

Undoubtedly, the participatory character of the German energy transition has enabled its success and progress, increasing the legitimacy of top-down measures (Fuchs & Wassermann, 2008; Wassermann et al., 2015). Another factor that certainly contributed to creating legitimacy was the technology development embedment in a community-scale approach and hence small- and medium-sized projects met greater local acceptance.

A current example of the importance of meaningful participation are measures for the national grid extension, which saw little progress while facing increasing protest and opposition in recent years. Neukirch (2017) found that frustration and opposition became widespread because community complaints (often) address systemic questions – challenging the grid extension as such when actually an alternative decentralized energy future appears possible, which is well beyond the scope of responsible authorities (Neukirch, 2020). Despite these challenges, studies confirm the importance of communication and public participation for preventing and managing conflicts in grid expansion (Kamlage et al., 2020). Here, fair and open processes that are not primarily characterized by the demand for the fastest and most cost-efficient realization of the project are essential (ibid).

Thus, in the recent past, the paradigm of participation has not lost its appeal but its efficacy to legitimize the energy transition is shaken. Although there is little doubt that participation is a prerequisite and essential tool for many transition measures (decision theatre, info fairs, dialogue forum, benefit-sharing and local ownership structures, etc.) (Nanz & Fritsche, 2012; Radke & Renn, 2019; Richter et al., 2016), policy measures no longer address the procedural and outcome dimension of participation. Instead, increasing criticism about persistent institutional forces, vested interests and top-down structures point to some of the shortfalls of participatory practices.

3.2. Renewable energy support that furthers participation: gaining output legitimacy

The other element enabling the energy transition in Germany were policy interventions, new regulations and planning regime that paved the way for integrating new technologies into the existing system (Feed in Law) and fostered an economic level playing field with financial incentives such as feed-in-tariffs (FiT). Small and collective actors were the main beneficiaries of these measures, since the incumbent players (RWE, EON and Vattenfall), disregarded and dismissed this emerging sector for decades. The *Renewable Energy Acts* until 2014 provided for a differentiated feed-in-tariff for every kWh of renewably generated electricity fed into the electricity grid, guaranteed for 20 years. This guaranteed rate of return was crucial in providing an equal playing field for smaller, community-sized renewable energy investors, providing much-needed investment security (Held et al., 2006).

Multi-partisan support for the FiT provided for in the Renewable Energy Act (EEG) provided a stable economic environment for the technology deployment and development of the sector. Indeed, the SPD and Greens government was succeeded by a CDU/CSU and SPD coalition which continued the support for renewable energy and further cemented the technology rise with the ultimate phase out of nuclear and later coal in the 2010s. The general public mood in Germany was very supportive of renewable energy deployment. Poll after poll confirmed support for strong climate goals and the energy transition is supported by the majority in the country. For example, in 2015, a representative survey found that 93% consider the increased expansion of renewable energies to be 'important' to 'extremely important' (AEE, 2015).

In addition, the regional planning regime played a role in legitimizing local energy projects early on by involving state, regional and local stakeholders in determining potential areas or preferential zones for renewable energy projects (specifically wind energy) (Müller & Morton, 2021; Wunderlich, 2012). These regional plans are also provided for public consultation with the purpose to identify points of conflict and local contention (Ziekow, 2018). Although the ultimate impact of the planning regime on local acceptance is questionable (Wunderlich, 2012), it has provided an impetus for many local governments to facilitate local initiatives for, e.g. community owned wind projects (AEE, 2012, 2013).

However, the criticism about the planning procedures has increased in recent years. On the one hand, the complexity of the planning regime embedded in multi-level governance system and the other hand technology advancements (growing height of the turbines) and in recent years has eroded support for wind energy projects and contributed to negative feelings being 'perceived as an encroaching, threatening presence in the landscape' (Müller & Morton, 2021, p. 70).

3.3. Marketization of renewables support – a danger to the legitimacy of the transition

However, policy changes towards a market-oriented and cost-centric policy approach benefitting merely large-scale deployment (incumbent actors) ultimately heralded a decline in the community energy sector. The EEG revision in 2012 significantly decreased the FiT scheme for renewable energy generators creating uncertainty among smaller investors (Smith Stegen & Seel, 2013). Major EEG reforms in June 2014 and amendments in 2016 moved the FiT scheme towards an auction approach fostering only large-scale renewable energy developments to be conducted by large developers and utilities while systematically marginalizing small- and medium-sized community actors (BBEn, 2015a, 2015b; BWE, 2016; Greenpeace, 2016; NABU, 2016). The auction mechanism increased the complexity of, and posed new hurdles for, community project developments (Maly et al., 2014; Tiedemann et al., 2015).

In fact, incumbent actors were able to take advantage of the new policy mechanism which became evident in the success of large bidders, including incumbent actors EON and EnBW (BBEn, 2015a). The first five auctions conducted in 2016 accepted a total of 603 bids for free-standing solar PV systems. An analysis of Bundesnetzagentur data from the first auctions showed that in the highly competitive process, only 2.15% of the total bids were coming from energy cooperatives, while from 144 successful bidders in five rounds, only two were energy cooperatives and 14 small- to medium-sized renewable energy projects (Mey, 2017).

This loss of actor diversity in Germany for the purpose of scale and cost undoubtedly influenced the support for renewable energy locally (Hauser et al., 2014). The unsolved complexity of federal and state responsibilities (e.g. planning laws for wind energy) further exacerbated public frustration and disappointment. Huge and Rossnagel (2018) study points to a number of deficits, for example, public participation in wind energy projects is often too late and only when the project is already in construction. They further find that communities have difficulties to know when and what engagement opportunities exist and less tangible issues of place attachment and concerns around landscape change and loss of property value are not considered in the formal approval procedures.

In addition, since the mid-2010s, the management of the energy transition has become more and more a point of criticism, with the public discourse lamenting the rising electricity costs and stagnating renewable energy deployment (in particular, planning of wind energy and grid extension) (BVerfG, 2021; Neukirch, 2017; Radtke, 2020). The annual German Social Sustainability Barometer (Wolf et al., 2021), provides a nuanced picture, revealing that a majority believe the costs and benefits of energy policy measures are unfairly distributed between households and companies. In the survey analysis, the authors conclude 'the transformation of the energy sectors can only succeed as a collective effort' this requires to also 'enable citizens to invest in the process' (Huge and Rossnagel 2018).

Of course, even before, participation was not all-inclusive (ability to involve all), nor created an immediate direct impact (decision to be influenced) or allowed for a fairness of the process on all accounts (Radtke, 2020; Radtke & Renn, 2019; Rommel et al., 2016). For example (Radtke, 2016) highlighted that community energy actors are primarily people from socioeconomically better-off cohorts of the population, primarily white men. In addition, many citizen energy projects are implemented by only a fraction of local residents, increasing potential social divisions. Nonetheless, local community ownership enabled local decision-making in the planning process and created a higher identification with local renewable energy assets. Beyond the procedural dimension, local ownership also addresses the distributional aspect since it empowers community members to financially benefit from the projects. Hence, the community ownership aspect was a powerful instrument to increase public legitimacy. Yet, ever-growing project sizes (costs and turbine heights) and the shift in the incentive scheme (from FiT to auctions) after the 2015 has left out smaller actors making community ownership more difficult.

The German Energiewende demonstrates that policy choices can lead to more or less meaningful participation in the transition process. Germany was particularly successful in the earlier stage of the transition to create public support through two mechanisms: sharing of benefits and involvement in decision-making processes: individuals and communities financially profited through ownership and participated in decision-making as part of the local planning process. Studies have shown that community-owned wind or solar farms (more often than not) achieved win-win situations for the locals for example in terms of regional added value (local jobs, investments etc.) (Gottschalk et al., 2016). Hence, allowing communities to become actors in the energy transitions can legitimize not only the input, but also the output of an energy transition. Although participation is widely and rightly acknowledged as essential for creating legitimacy (Rau et al., 2012; Schweizer-Ries & Renn, 2013), Germany risks its achievements since community ownership has become more unattainable and deficits of public engagement in the legislated planning processes hamper local acceptance for energy projects.

4. Legitimizing the energy transition in Australia?

In comparison to Germany, citizen energy and community energy projects are still playing a minor role in Australia's transition – especially renewable energy support schemes that have mostly supported large commercial industry participants over community-scale projects. However, as both mitigation pressures, and especially, adaptation pressures on infrastructure are rising, Australian state and federal governments are considering how community involvement can be enhanced. This includes community consideration for decision-making on the location and choice of energy sources and supporting more active community participation and decentralized energy projects.

Australia is a country rich in energy resources: both renewable and fossil fuel resources are plentiful. In recent years, Australia has been rapidly adding renewable energy capacity to its electricity system, mostly large-scale wind generation. Wind provides over 35% of renewable generation in Australia (CEC, 2021). In addition, increasingly household-scale solar power is being added (23.5% of renewable generation, CEC, 2021), with one in three Australian households having rooftop solar panels (ARENA, 2021).

Australia's traditional energy generation, predominantly coal-powered, is now under severe economic pressure, with 12 of 34 coal-fired power plants closed in the last 10 years (Burke et al., 2018) and further closures already announced (CEC, 2021). Renewable energy is slated to replace the lost capacity – but will also require a reconfiguration of networks. These infrastructural changes to how energy is generated and distributed are already having an impact on local communities. In addition, climate impacts with increased risk of bushfire, inundation and flooding are impacting the Australian energy system and decentralized, regional energy solutions relying on renewable sources are increasingly considered. Gaining and maintaining legitimacy of the transition will be central to its ongoing success.

4.1. Focussing on the large scale – a transition without participation?

The economic reliance on cheap fossil fuels has seen conservative governments remain extremely reluctant to commit to energy transition efforts. Initially, climate policy in Australia, therefore, focused on a 'no regrets' approach, where voluntary action was preferred over prescriptive policy measures (Bulkeley, 2001), as climate measures should not come at a cost to the economy. Australia, while a founding member of the UNFCCC, only ratified the Kyoto Protocol in 2007, even after negotiating a very favourable target, which allowed Australia to actually increase emissions above the 1990 baseline scenario.

From 1995 onwards, Australian states developed a number of state-level climate mitigation schemes. These included, for example, the New South Wales Greenhouse Gas Abatement Scheme and the ACT Greenhouse Gas Reduction Scheme. Following increasing public demand for climate action, the federal government followed suit and implemented a limited federal renewable energy scheme in 2000. The initial scheme required 2% of electricity to be sourced from renewable energy sources by 2010. This target was achieved well ahead of time and initially not extended by the government. Public pressure to address climate change played a central role in the change of government in 2007, when a labour government under Kevin Rudd was elected (Curran, 2011; Rootes, 2008). The Rudd government not only ratified the Kyoto Protocol and attempted to introduce an emissions trading scheme, but also substantially expanded the Renewable Energy Target (RET) to a 20% goal.

The Australian RET seeks to incentivize the lowest cost abatement or lowest cost renewables. The RET relies on requiring large users or retailers to buy a set percentage of renewable energy. Unlike the German FIT scheme, target schemes such as the RET support the lowest-cost renewable energy, which, in Australia, has initially been large-scale wind. In 2010, the RET was split into a large-scale and a small-scale target. The small-scale renewable energy scheme functions as a feed-in-tariff with a fixed price, but only includes a small part of the target (less than 10% of the overall capacity the RET sought to achieve) (Buckman & Diesendorf, 2010; Valentine, 2010).

As a result, siting decisions were made based on resource availability and proximity to the network – a development that is now well recognized as suboptimal and in need of reform (e.g. ESB, 2022). Some locations close to the electricity transmission network, high wind resources, and low population density have seen considerable over-development. For example, in the shire of Moyne, in the state of Victoria, 14% of the land mass is now occupied by wind power sites and electricity transmission infrastructure. Poor community project development, community engagement practices and lack of community benefit from the industrialization of the landscape have created tension between local communities and further renewable development (Moyne Shire Council, 2019). Loss of legitimacy associated with this development is well described in the literature (Fuchs, 2019; Genus & Iskandarova, 2020; Hindmarsh, 2010; Markard et al., 2016; Ruef & Markard, 2010). In Moyne, the council is currently developing a wind farm position that strongly calls for more and better community participation and benefit-sharing, clearly expressing discontent with the current planning law framework (Moyne Shire Council, 2022). This includes calls for buffer zones, caps on wind farm developments, undergrounding of power lines and benefits for hosting communities.

Planning law has a central role to play in seeking to balance transition pressures with local interests. (Caripis & Kallies, 2012). Overall, planning laws around the country have become more inclusive of local concerns. In the state of Victoria for example, the relevant regulations contained in each planning scheme for each local government require a planning authority assessing a renewable energy development to take into account relevant guidelines, including community engagement and benefit-sharing guidelines (Victorian Government, 2021c) and more specific guidelines for solar energy facilities (Victorian Government, 2019) and Wind Energy Facilities (Victorian Government, 2021d), both of which require extensive and early community engagement by proponents, a lack of which impacting their chances for approval. Formalized opportunities for submissions and objections of affected individuals to planning decisions are available through the states' planning law frameworks. To expand on the Victorian example, a veto right for owners of dwellings within a certain distance of new wind farm developments has arguably led to a need to engage impacted communities earlier. As a further, 'soft' option to address conflict, the federal government created the Office of the National Wind Farm Commissioner in 2015. Now rebadged as Australian Energy Infrastructure Commissioner (AEIC), it provides inter alia a voluntary process for conflict resolution between proponents and objectors at all stages of a project; proactively engages with industry and stakeholders; and seeks to further best practice approaches for community engagement (see AEIC, 2021). A series of recommendations by the Office ultimately support a more consultative style of decision-making on siting of large energy infrastructure – including not only wind, but storage facilities, solar installations and network infrastructure.

State governments have likewise shifted renewable support instruments to expressly require community engagement. Both the ACT and Victoria have introduced reverse auction schemes, which require successful applicants to purposefully engage and benefit-share with the community in which their projects are located (Buckman et al., 2014; Victorian Government, 2021c). While these changes can support better community engagement, they do not support community energy, which arguably was central to the German transition.

4.2. Supporting households – the individualization of the energy transition

However, the Australian energy transition is not only characterized by large-scale wind and solar developments. Australia is also a world leader in rooftop solar by owner-occupiers. This development was driven early on by the choice of support mechanisms. From 2008 onwards, all Australian states and territories introduced feed-in tariffs for household-scale renewable installations. Feed-in-rates were initially set comparatively high and installation rates went rapidly up. While all schemes were considerably cut back within years, the trend continued, with the federal small-scale renewable energy target providing support, but also technology costs falling.

Rooftop solar does give many Australian households the opportunity to participate in the energy transition as actors. However, it is not readily available to low-income earners, renters or generally apartment dwellers (ACOSS, BSL & TCI, 2017; Rutovitz et al., 2018), and considerably impacts the equity of the energy transition (Best et al., 2021). Australian support for rooftop solar focuses on the individual household rather than the community. While local governments have been providing bulk-buy schemes that have played an important role in lowering investment costs, investment decisions are the prerogative of the individual homeowner (Hadfield & Cook, 2018). This does not only exclude considerable parts of the population from access to the benefit of rooftop solar, but the large uptake of small-scale solar energy is now also negatively impacting on the shared network (McConnell, 2021). Network costs are paid by all energy consumers – whether they own solar rooftop or not – including network augmentation necessary to facilitate the two-way flow of energy from prosumers to the network. In other words, solar rooftop owners do have the benefits of a lower electricity bill but share the costs they are causing to the system.

Overall, due to the design of support mechanisms, renewables investment in Australia clusters either on the large-scale or individualized household-scale ends of the system. Neither is particularly conducive to the type of community participation and buy-in that characterized the German transition effort.

4.3. An opportunity for participation and creating output legitimacy

What is notable in the Australian context is a lack of medium-sized community-owned renewable support – incidentally exactly the type of installation on which Germany's success was built. This is not to say that community energy projects do not exist in Australia. On the contrary, there is a diverse and vibrant community energy sector with now more than 150 projects operating (Community Power Agency, 2021). Nevertheless, community energy remains a niche operator in Australia, and existing projects have arguably happened despite and not because of policy support (Mey & Hicks, 2019). However, the attention of policymakers, both federal and state, is now shifting to consider support for a range of schemes that rely on community involvement. Especially notable is the shift from community involvement in generation to community involvement in managing energy transitions – driven not only by the need for just transitions, but also because of the promise of a more resilient distributed energy system in the face of climate change.

Examples include support for microgrid and edge-of-grid developments (Australian Government, 2020; Victorian Government, 2017, 2020), neighbourhood batteries (Victorian Government, 2021a), and solar support for apartment buildings and community housing (Victorian Government, 2017; Victorian Government, 2021b), all of which require coordination and commitment to sharing between participating members or communities. Successful implementation of these types of mechanisms can support genuine community buy-in for the energy transition and thereby enhance output legitimacy. Developing policy instruments that can move these projects from pilot projects to become a standard of Australia's energy transition could help to bridge the gap between industrialized large-scale and individualized small-scale developments.

5. Discussion and conclusion

In this article, we argue that in particular the output legitimacy created through community participation is a prerequisite for successful energy transitions. To this purpose, we drew on the examples of Germany and Australia to illustrate how policy instruments can play a decisive role in enabling participation at different stages of the transition process. While we acknowledge that the case studies show high individual differences in the way in which transitions are approached and supported, we can nevertheless see patterns emerging of how community participation can enhance output legitimacy. Two main 'battlegrounds' for community participation emerge from both case studies, including firstly, the scope legal frameworks allow for community as an active actor, investing in and benefitting from the transition and secondly, the role of planning frameworks in channelling communities' discontent.

In Germany, early legal frameworks for renewable energy policies and land use planning enabled citizen and community participation across the process and outcome dimensions. The FiT policy created a more level playing field and incentivized community ownership and involvement in decision-making as part of the local planning process. The combination of different top-down policy measures and bottom-up initiatives led to a beneficial environment in the first decades of the energy transition. But the German example also shows that legitimacy of an energy transition can be threatened when community buy-in is lost. While Germans remain overwhelmingly committed to the energy transition and, thus arguably, input legitimacy remains high, the same cannot be said for output legitimacy. Loss of opportunity for community participation in the energy transition, materializes, in particular, in delays in planning processes. Rising opposition towards large-scale generation and transmission projects, leads to longer durations of planning and approval procedures, slowing down the national transition efforts significantly. Since most new developments are no longer community wind farms or solar parks, but rather investments by large corporate actors, financial benefits for a community of locality or community of interest are less likely and hence resentment and feeling of inequitable distribution of costs and benefits have risen.

In contrast, Australia's energy transition started much later with already more advanced technologies (size and capital costs) and a stronger market-oriented approach excluding small and medium-sized collective actors. Hence, Australians were to some extent, deprived of the opportunity for community ownership (and its financial benefits) and involvement in decision-making processes at a local level. In addition, planning laws did not consider local involvement for a long time, and despite improvements, is only slowly building adequate procedures and conditions for community participation and benefit-sharing. Overwhelming support for the energy transition and more than three million rooftop-solar households confirm that Australia's climate and energy policies have high input legitimacy. But, drawing on the German experience, output legitimacy remains on shaky ground, as large parts of the population (such as renters and low-income households) are still excluded from individual participation.

There is clearly a large appetite for more meaningful and inclusive participation and community-scale projects have seen a revival in the last years. Initiatives and concepts such as community microgrids (the current focus on energy communities at the edge of the grid or even off the main grid), solar gardens or solar banks (behind and in front of the meter) (Rutovitz et al., 2018; Haystacks, 2022; Enova, 2020) have gained traction in the last years and can provide an opportunity for a more inclusive, and thereby more legitimate transition.

For both jurisdiction, community involvement through planning laws appears as too late in the process – becoming an outlet to express dissatisfaction. Arguably, meaningful participation is required much earlier (at the concept stage) through opportunities for influencing decision-making and co-ownership (Hicks et al. 2018). Early examples of trying to include community benefit-sharing as part of large-scale investment are emerging in Australian state planning frameworks, and the outcomes of these requirements warrant careful observation.

Both the German and the Australian energy transition are now accelerating. In Germany, this is due to the Ukraine crisis and the associated need to wean the country off its dependence on Russian gas. The 'Oster-Paket' (Easter package) of energy law reforms (BMWK, 2022) includes ambitious targets to speed up the transition. Although the paket commits to lower the hurdles for community ownership by enabling them outside of the auction scheme (this applies to wind projects up to 18 MW and solar projects up to 6 MW). For the sake of speed, planning frameworks and approval procedures will be streamlined with (potentially) negative consequences for participation (less time).

In Australia, the recent change of federal government sees renewed commitment and legislative activity to support a rapid transition. Labor plans to support the set up of more than 80 'solar banks' – community-owned solar projects designed to break down the barriers to solar ownership faced by one-third of Australian households, such as renters and apartment dwellers. In addition, the installation of 400 community batteries projects which will unlock new community energy ownership models that seek to assist groups (such as communities, households or businesses) access clean energy technologies, set up a community energy retailer, or scale-up generation and storage capabilities (ALP, 2022).

Germany appears to continue losing on the dimension of output legitimacy, while recent policy efforts at least show an increased awareness of its importance. However, it remains to be seen if participation is not traded for speed which will ultimately thwart the energy transition efforts. The same applies to Australia, where the government gives more weight to the output dimension of legitimacy in recent policy announcements. After all, the experience detailed in this paper shows that rapid transitions without output legitimacy risk the sustained success of the transition effort.

Decision-makers would be well advised to consider both – the input and output legitimacy of the measures chosen to support the transition and take special care to support community participation as part of this process to ensure the lasting success of the current transition effort. After all, meaningful involvement of citizens matters which requires political will and courage.

Disclosure statement

No potential conflict of interest was reported by the author(s).

ORCID

Franziska Mey http://orcid.org/0000-0002-3401-3235

References

ACOSS, BSL, & TCI. (2017). *Empowering disadvantaged households to access affordable, clean energy.* Australian Council of Social Service, Brotherhood of St Laurence, The Climate Institute.

AEE. (2012). KOMM:MAG — Das Jahresmagazin zu Erneuerbaren Energien in Kommunen [The annual magazine about renewable energy in local communities]. *Magazin by Agentur Fuer Erneuerbare Energien, 39*, 25–35.

AEE. (2013). *Bürgerbeteiligung. Die Energiewende gestalten* [Civic participation. Forming the energy transition] *(Vol. 2). Agentur für Erneuerbare Energien.* https://www.unendlich-viel-energie.de/mediathek/publikationen/kommmag-das-jahresmagazin-zu-erneuerbaren-energien-in-kommunen-201213

AEE. (2015). Akzeptanzumfrage Erneuerbare Energien 2015. *Agentur für Erneuerbare Energien.* Retrieved October 23, 2022, from https://www.unendlich-viel-energie.de/mediathek/grafiken/umfrage-akzeptanz-erneuerbare-energien-2015

AEE. (2022). Akzeptanzumfrage Erneuerbare Energie 2021. *Agentur für Erneuerbare Energien.* Retrieved January 30, 2022, from https://www.unendlich-vielenergie.de/mediathek/grafiken/aee-akzeptanzumfrage-2021

AEIC. (2021). Australian Energy Infrastructure Commissioner. https://www.aeic.gov.au/about#what-we-do

Aldrich, H. E., & Fiol, C. M. (2007). Fools rush in? The institutional context of industry creation. *Entrepreneurship: Concepts, Theory and Perspective, 19*(4), 105–127. https://doi.org/10.1007/978-3-540-48543-8_5

ALP (2022). Website. Retrieved on 10th February 2022 from https://www.alp.org.au/

ARENA. (2021). *Solar energy.* https://arena.gov.au/renewable-energy/solar/>

Arnstein, S. R. (1969). A ladder of citizen participation. *Journal of the American Institute of Planners, 35*(4), 216–224. https://doi.org/10.1080/01944366908977225

Australian Government. (2020). Department of industry, science, energy and resources, regional and remote communities reliability fund - Microgrids 2020–21.

BBEn. (2015a). Pressemitteilung Bündnis Bürgerenergie kritisiert Verdrängung kleiner Bieter [Press release Alliance Citizen Energy criticizes surpressions of smaller bidders]. Retrieved September 14, 2015, from https://www.buendnis-buergerenergie.de/buergerenergie/definition/

BBEn. (2015b). *Stellungnahme des Bündnis Bürgerenergie zum Eckpunkte Papier "Ausschreibungen für die Förderung von Erneuerbaren Energien"* [Statement of the Alliance Citizen Energy in relation to the cornerstone paper "tender for funding renewable energy"]. http://www.buendnis-buergerenergie.de/fileadmin/user_upload/downloads/Positionspapiere/BBEn_Stellungnahme_Ausschreibungen_von_PV-Freiflaechenanlagen.pdf

Bergek, A., Jacobsson, S., & Sandén, B. A. (2008). "Legitimation" and "development of positive externalities": two key processes in the formation phase of technological innovation systems. *Technology Analysis & Strategic Management, 20*(5), 575–592. https://doi.org/10.1080/09537320802292768

Best, R., Chareunsy, A., & Li, H. (2021). Equity and effectiveness of Australian small-scale solar schemes. *Ecological Economics, 180*, 106890. https://doi.org/10.1016/j.ecolecon.2020.106890

BMWi. (2021). Unsere Energiewende: sicher, sauber, bezahlbar. Retrieved October 29, 2021, from https://www.bmwi.de/Redaktion/DE/Dossier/energiewende.html

BMWK. (2022). Bundesministerium für Wirtschaft und Klimaschutz: Überblickspapier Osterpaket (6 April 2022). https://www.bmwk.de/Redaktion/DE/Downloads/Energie/0406_ueberblickspapier_osterpaket.pdf

Buckman, G., & Diesendorf, M. (2010). Design limitations in Australian renewable electricity policies. *Energy Policy, 38*(7), 3365–3376. https://doi.org/10.1016/j.enpol.2010.02.009

Buckman, G., Sibley, J., & Bourne, R. (2014). The large-scale solar feed-in tariff reverse auction in the Australian capital territory, Australia. *Energy Policy, 72*, 14–22. https://doi.org/10.1016/j.enpol.2014.04.007

Bulkeley, H. (2001). No regrets?: Economy and environment in Australia's domestic climate change policy process. *Global Environmental Change, 11*(2), 155–169. https://doi.org/10.1016/S0959-3780(00)00064-9.

Burke, P. J., Best, R., & Jotzo, F. (2018). Closures of coal-fired power stations in Australia: Local unemployment effects (No. 1809; CCEP Working Paper).

BVerfG. (2021). Bundesverfassungsgericht (German Constitutional Court), 'Constitutional Complaints against the Federal Climate Change Act partially successful' (Media Release no. 31/2021, 29 April 2021), https://www.bundesverfassungsgericht.de/SharedDocs/Pressemitteilungen/EN/2021/bvg21-031.html

BWE. (2016). *Stellungnahme zum Entwurf eines Gesetzes zur Einführung von Ausschreibun- gen für Strom aus erneuerbaren Energien und zu weiteren Änderungen des Rechts der erneuerbaren Energien (Erneuerbare-Energien-Gesetz 2016) BT-Drs. 18/8860.* Berlin. www.wind-energie.de

Caripis, L., & Kallies, A. (2012). "Planning away" Victoria's renewable energy future? Resolving the tension between the local and global in windfarm developments. *Environmental and Planning Law Journal, 29*(5), 415–435. https://researchrepository.rmit.edu.au/esploro/outputs/journalArticle/Planning-away--Victorias-renewable-energy/9921862194501341#details

Clean Energy Council. (2021). *Clean Energy Australia Report 2021.* Melbourne. http://www.cleanenergycouncil.org.au/policy-advocacy/reports.html

Coenen, L., Campbell, S., & Wiseman, J. (2018). Regional innovation systems and transformative dynamics: Transitions in coal regions in Australia and Germany BT – New avenues for regional innovation systems - theoretical advances, empirical cases and policy lessons. In A. Isaksen, R. Martin, & M. Trippl (Eds.), (pp. 199–217). Springer International Publishing. https://doi.org/10.1007/978-3-319-71661-9_10

Community Power Agency. (2021). Personal Correpondence on the number of Community Energy Projects in Australia.

Cowell, R., Bristow, G., & Munday, M. (2011). Acceptance, acceptability and environmental justice: The role of community benefits in wind energy development. *Journal of Environmental Planning and Management, 54*(4), 539–557. https://doi.org/10.1080/09640568.2010.521047.

Creamer, E., Taylor Aiken, G., van Veelen, B., Walker, G., & Devine-Wright, P. (2019). Community renewable energy: What does it do? Walker and Devine-Wright (2008) ten years on. *Energy Research & Social Science, 57*(August), 101223. https://doi.org/10.1016/j.erss.2019.101223.

Curran, G. (2011). Modernising climate policy in Australia: Climate narratives and the undoing of a Prime Minister. *Environment and Planning C: Government and Policy, 29*(6), 1004–1017. https://doi.org/10.1068/c10217

Deutscher Bundestag. (1990). Gesetzes über die Einspeisung von Strom aus erneuerbaren Energien in das öffentliche Netz (Stromeinspeisungsgesetz) [Act about the feed in of power from renewable energy into the public grid (Feed in Law)], 11/7816 (1990).

Deutscher, Bundestag. (2000). Gesetz für den Vorrang Erneuerbarer Energien (Erneuerbare-Energien-Gesetz – EEG) sowie zur Änderung des Energiewirtschaftsgesetzes und des Mineralölsteuergesetzes.

DGRV. (2016). DGRV-Jahresumfrage unter Energiegenossenschaften: Gründungszahlen weiter rückläufig [DGRV-annual survey amongs the energy cooperatives: founding numbers continuous to decline]. Retrieved August 17, 2016, from http://www.genossenschaften.de/dgrv-jahresumfrage-unter-energiegenossenschaften-gr-ndungszahlen-weiter-r-ckl-ufig

Energy Enova. (2020). Website. Retrieved on 21st August 2020 from www.enovaenergy.com.au.

ESB. (2022). Transmission Access Reform Consultation Paper (May 2022). https://www.energy.gov.au/sites/default/files/2022-06/Transmission%20access%20reform%20Consultation%20paper.pdf

Fuchs, G. (2019). Legitimacy and field development: Electricity transition(s) in Germany. *Global Transitions, 1*, 141–147. https://doi.org/10.1016/j.glt.2019.10.001

Fuchs, G., & Wassermann, S. (2008). Picking a winner? Innovation in photovoltaics and the political creation of Niche Markets Gerhard. *Science, Technology & Innovation Studies, 4*(2), 94–113.

Genus, A., & Iskandarova, M. (2020). Transforming the energy system? Technology and organisational legitimacy and the institutionalisation of community renewable energy. *Renewable and Sustainable Energy Reviews, 125*(January), 109795. https://doi.org/10.1016/j.rser.2020.109795

Gottschalk, M., Hoppenbrock, C., Kucharczak, L., Schäfer, S., & Wetzel, H. (2016). *Regionale Wertschöpfung in der Windindustrie am Beispiel Nordhessen.* Kassel.

Greenpeace. (2016). *Greenpeace-Stellungnahme zum Erneuerbare-Energien- Gesetz 2016* [Greenpeace position statement regarding the renewable energy].

Gross, C. (2007). Community perspectives of wind energy in Australia: The application of a justice and community fairness framework to increase social acceptance. *Energy Policy, 35*(5), 2727–2736. https://doi.org/10.1016/j.enpol.2006.12.013

Hadfield, P., & Cook, N. (2018). Financing the low-carbon city: Can local government leveragePublic finance to facilitate equitable decarbonisation?. *Urban Policy and Research*. https://doi.org/10.1080/08111146.207.1421532.

Harrahill, K., & Douglas, O. (2019). Framework development for "just transition" in coal producing jurisdictions. *Energy Policy*, https://doi.org/10.1016/j.enpol.2019.110990

Harris-Lovett, S. R., Binz, C., Sedlak, D. L., Kiparsky, M., & Truffer, B. (2015). Beyond user acceptance: A legitimacy framework for potable water reuse in California. *Environmental Science & Technology*, *49* (13), 7552–7561. https://doi.org/10.1021/acs.est.5b00504

Hauser, E., Weber, A., Zipp, A., & Leprich, U. (2014). *Bewertung von Ausschreibungsverfahren als Finanzierungsmodell für Anlagen erneuerbarer Energienutzung* [Evaluation of the tender process for the funding modell of the renewable energy systems]. http://www.bee-ev.de/Publikationen/IZES2014-05-20BEE_EE-Ausschreibungen_Endbericht.pdf

Heffron, R. J., & McCauley, D. (2017). The concept of energy justice across the disciplines. *Energy Policy*, *105* (November 2016), 658–667. https://doi.org/10.1016/j.enpol.2017.03.018

Held, A., Ragwitz, M., & Haas, R. (2006). On the success of policy strategies for the promotion of electricity from renewable energy sources in the EU. *Energy & Environment*, *17*(6), 849–868. https://doi.org/10.1260/095830506779398849.

Hicks, J., & Ison, N. (2018). An exploration of the boundaries of 'community' in community renewable energy projects: Navigating between motivations and context. *Energy Policy*, *113*(June 2016), 523–534. https://doi.org/10.1016/j.enpol.2017.10.031.

Hicks, J., Lane, T., Wood, E., & Hall, N. (2018). Enhancing positive social outcomes from wind farm development: Evaluating community engagement and benefit-sharing in Australia. *Clean Energy Council.* Retrieve on 2nd October 2021 from https://assets.cleanenergycouncil.org.au/documents/resources/reports/enhancing-positive-social-outcomes-from-wind-farmdevelopment. pdf

Hildebrand, J., Rau, I., & Schweizer-Ries, P. (2018). Akzeptanz und Beteiligung – ein ungleiches Paar. In L. Holstenkamp, & J. Radtke (Eds.), *Handbuch Energiewende und Partizipation* (pp. 195–209). Springer Fachmedien Wiesbaden. https://doi.org/10.1007/978-3-658-09416-4_12

Hindmarsh, R. (2010). Wind farms and community engagement in Australia: A critical analysis for policy learning. *East Asian Science, Technology and Society: An International Journal*, *4*(4), 541–563. https://doi.org/10.1007/s12280-010-9155-9

Holstenkamp, L., & Radtke, J. (2018). *Handbuch Energiewende und Partizipation.* Springer VS.

Huge, A., & Roßnagel, A. (2018). Möglichkeiten der Öffentlichkeitsbeteiligung in Planungs- und Genehmigungsverfahren von Windenergieanlagen BT - Handbuch Energiewende und Partizipation. In L. Holstenkamp & J. Radtke (Eds.), (pp. 613–625). Springer Fachmedien Wiesbaden.

Haystacks Solar Gardens Website. (2022). Retrieved on 22nd October 2022 from https://haystacks.solargarden.org.au/.

Jenkins, K., Hurlbert, M., Rayner, J., Mundaca, L., Busch, H., Schwer, S., Bouzarovski, S., Simcock, N., Sovacool, B. K., Burke, M., Baker, L., Kotikalapudi, C. K., & Wlokas, H. (2018). "Successful" low-carbon energy transitions at the community level? An energy justice perspective. *Applied Energy*, *218*(October 2017), 292–303. https://doi.org/10.1016/j.erss.2017.11.015

Jenkins, K., McCauley, D., & Forman, A. (2017). Energy justice: A policy approach. *Energy Policy*, *105*, 631–634. https://doi.org/10.1016/j.enpol.2017.01.052

Kahla, F., Holstenkamp, L., Müller, J. R., & Degenhart, H. (2017). *Entwicklung und Stand von Bürgerenergiegesellschaften und Energiegenossenschaften in Deutschland.* Lüneburg.

Kamlage, J.-H., Drewing, E., Reinermann, J. L., Vries, N. D., & Flores, M. (2020). Fighting fruitfully? Participation and conflict in the context of electricity grid extension in Germany. *Utilities Policy*, *64* (January), https://doi.org/10.1016/j.jup.2020.101022

Kamlage, J.-H, Nanz, P., & Fleischer, T. (2014). Bürgerbeteiligung und Energiewende: Dialogorientierte Bürgerbeteiligung im. In H.-C Rogall et al. (Eds.), *4. Jahrbuch Nachhaltige Ökonomie. Im Brennpunkt: Die Energiewende als gesellschaftlicher Transformationsprozess* (pp. 196–216). Metropolis Verlag.

MacArthur, J. L. (2016). Challenging public engagement: Participation, deliberation and power in renewable energy policy. *Journal of Environmental Studies and Sciences*, *6*(3), 631–640. https://doi.org/10.1007/s13412-015-0328-7

Maly, C., Meister, M., & Schomerus, T. (2014). EEG 2014 – Das Ende der Bürgerenergie? [Renewable energy Act 2014 - the end of the citizen energy?]. *EnergieRecht Zeitschrift Für Die Gesamte Energierechtspraxis*, (4), 147–154. https://www.erdigital.de/ER.04.2014.147

Markard, J., Wirth, S., & Truffer, B. (2016). Institutional dynamics and technology legitimacy - A framework and a case study on biogas technology. *Research Policy*, *45*(1), 330–344. https://doi.org/10.1016/j.respol.2015.10.009

Mautz, R., Byzio, A., & Rosenbaum, W. (2008). *Auf dem Weg zur Energiewende Die Entwicklung der Stromproduktion aus erneuerbaren Energien in Deutschland* [On the way to the energy transition the Development of power generation from renewable energies in Germany]. Universitätsverlag Göttingen. http://webdoc.sub.gwdg.de/univerlag/2008/energiewende.pdf

McConnell. (2021). Solar Justice: Who should pay for the Grid? (A BC, 7 Jun 2021) https://www.abc.net.au/religion/solar-justice-who-should-pay-for-the-grid/13376586

Mey, F. (2017). *Development of Community Renewable Energy and the Role of Institutional Factors: A Three Country Comparison (PhD thesis).* University of New South Wales.

Mey, F., & Hicks, J. (2019). Community owned renewable energy: Enabling the transition towards renewable energy? BT – Decarbonising the built environment: Charting the transition. In P. Newton, D. Prasad, A. Sproul, & S. White (Eds.), (pp. 65–82). Springer Singapore. https://doi.org/10.1007/978-981-13-7940-6_4

Morris, C., & Jungjohann, A. (2016). *Energy democracy Germany's Energiewende to renewables.* Palgrave Macmillan. https://doi.org/10.1007/978-3-319-31891-2

Moyne Shire Council. (2019). Council urges Planning Minister to Take Action on Wind Farms (Press Release, 5 April 2019) https://www.moyne.vic.gov.au/News-Media/Council-urges-Planning-Minster-to-take-action-on-wind-farms

Moyne Shire Council. (2022). Draft Wind Farm Position https://www.moyne.vic.gov.au/Your-Say/Draft-Wind-Farm-Position

Müller, K., & Morton, T. (2021). The space, the time, and the money. *Wind energy politics in East Germany, 40* (May), 62–72.

Mundaca, L., Busch, H., & Schwer, S. (2018). "Successful" low-carbon energy transitions at the community level? An energy justice perspective. *Applied Energy*, *218*(October 2017), 292–303. https://doi.org/10.1016/j.apenergy.2018.02.146

NABU. (2016). *NABU-Stellungnahme zum EEG- Referentenentwurf* [NABU position statement regarding the Renewable Energy Act].

Nanz, P., & Fritsche, M. (2012). *Handbuch Bürgerbeteiligung.* Bundeszentrale für politische Bildung.

Nanz, P., & Kamlage, J.-H. (2014). Bürgerbeteiligung und Energiewende : Dialogorientierte Bürgerbeteiligung im Jahrbuch Nachhaltige Ökonomie 2014 / 2015. In H. Rogall, H.-C. Binswanger, A. G. Felix Ekardt, W.-D. Hasenclever, I. Hauchler, M. Jänicke, K. Kollmann, N. V. Michaelis, & H. G. Nutzinger (Eds.), *Jahrbuch Nachhaltige Ökonomie 2014 / 2015* (pp. 195–216). Metropoli.

Neukirch, M. (2017). *Die Dynamik des Konflikts um den Stromtrassenbau: Stabilität, Wandel oder Stagnation?* (SOI Discussion Paper No. 2017–4). Stuttgart. http://hdl.handle.net/10419/170579%0AStandard-Nutzungsbedingungen

Neukirch, M. (2020). Grinding the grid: Contextualizing protest networks against energy transmission projects in Southern Germany Bavaria. *Energy Research & Social Science*, *69*(April). https://doi.org/10.1016/j.erss.2020.101585.

Radtke, J. (2016). *Bürgerenergie in Deutschland Partizipation zwischen Gemeinwohl und Rendite [Citizen energy in Germany participation between public interest and economic return].* Springer VS.

Radtke, J. (2020). Das Jahrhundertprojekt der Nachhaltigkeit am Scheideweg. *Zeitschrift Für Politikwissenschaft*, *30*(1), 97–111. https://doi.org/10.1007/s41358-020-00215-6

Radtke, J., & Renn, O. (2019). Partizipation und bürgerschaftliches Engagement in der Energiewende - Energiewende: Eine sozialwissenschaftliche Einführung. In J. Radtke, & W. Canzler (Eds.), (pp. 283–316). Springer Fachmedien Wiesbaden. https://doi.org/10.1007/978-3-658-26327-0_10

Rau, I., Hildebrand, J., & Weingarten, E. (2012). *Akzeptanz durch Beteiligungsverfahren beim Netzausbau: Möglichkeiten und Grenzen des rechtlichen Rahmens.* (UVP-Report No. 26).

REN21. (2021). *Renewables 2021 - Global status report. Global Status Report for Buildings and Construction: Towards a Zero-emission, Efficient and Resilient Buildings and Construction Sector.* Paris: REN21 Secretariat. https://www.ren21.net/wp-content/uploads/2019/05/gsr_2020_full_report_en.pdf%0Ahttp://www.ren21.net/resources/publications/

Richter, I., Danelzik, M., Molinengo, G., Nanz, P., & Rost, D. (2016). Bürgerbeteiligung in der Energiewende Zusammenfassung. Potsdam. https://publications.iass-potsdam.de/rest/items/item_1443976_3/component/file_1443980/content

Rommel, J., Radtke, J., von Jorck, G., Mey, F., & Yildiz, Ö. (2016). Community renewable energy at a crossroads: A think piece on degrowth, technology, and the democratization of the German energy system. *Journal of Cleaner Production*, https://doi.org/10.1016/j.jclepro.2016.11.114

Rootes, C. (2008). The first climate change election? The Australian general election of 24 November 2007. *Environmental Politics*, *17*(3), 473–480. https://doi.org/10.1080/09644010802065815

Ruef, A., & Markard, J. (2010). What happens after a hype? How changing expectations affected innovation activities in the case of stationary fuel cells. *Technology Analysis & Strategic Management*, *22*(3), 317–338. https://doi.org/10.1080/09537321003647354

Rutovitz, J., McIntosh, L., Ison, N., Noble, E., Hicks, J., & Mey, F. (2018). Social access solar gardens for Australia. Sydney. https://www.uts.edu.au/research-and-teaching/our-research/institute-sustainable-futures/our-research/energy-and-climate/social-solar-gardens

Sareen, S., & Haarstad, H. (2020). Legitimacy and accountability in the governance of sustainable energy transitions. *Global Transitions*, *2*, 47–50. https://doi.org/10.1016/j.glt.2020.02.001

Scharpf, F. W. (1999). *Governing in Europe: Effective and democratic?* Oxford University Press.

Scharpf, F. W. (2003). *Problem-Solving Effectiveness and Democratic Accountability in the EU* (MPIfG Working Paper 03/1 No. 03/1). Köln.

Schneider, D., & Boenigk, N. (2012). *Renews Spezial - Planungsrecht & Erneuerbare Energien* [Renews special - Planning law and renewable energies]. Agentur für Erneuerbare Energien e. V.

Schweizer, P.-J., & Renn, O. (2013). Partizipation in Technikkontroversen: Panakeia für die Energiewende?. *TATuP – Zeitschrift Für Technikfolgenabschätzung in Theorie Und Praxis*, *22*(2), 42–47. https://doi.org/https://doi.org/10.14512/tatup.22.2.42.

Schweizer-Ries, P., Hildebrand, J., Hinse, M., Rühmland, S., & Rau, I. (2016). *Begleit- und Akzeptanzforschung zu aktuellen Fragen des Stromnetzausbaus in Deutschland - Wissenschaftliche Begleitung der Planungspraxis (Akzeptanz Netzausbau)*. https://doi.org/10.2314/GBV:872386333

Smith Stegen, K., & Seel, M. (2013). The winds of change: How wind firms assess Germany's energy transition. *Energy Policy*, *61*, 1481–1489. https://doi.org/10.1016/j.enpol.2013.06.130

Snell, D. (2018). "Just transition"? conceptual challenges meet stark reality in a "transitioning" coal region in Australia. *Globalizations*, *15*(4), 550–564. https://doi.org/10.1080/14747731.2018.1454679

The Australia Institute. (2021). *Climate of the Nation 2021*. Canberra. https://australiainstitute.org.au/wp-content/uploads/2021/10/211013-Climate-of-the-Nation-2021-WEB.pdf

Tiedemann, S., Wigand, F., & Klessmann, C. (2015). *Akteursvielfalt Windenergie an Land. Herausforderungen, Akteursdefinition und mögliche Sonderregelungen* [Actors diversity wind energy in rural areas. Challenges, actor definitions and possible special regulations].

Valentine, S. (2010). Braking wind in Australia: A critical evaluation of the renewable energy target. *Energy Policy*, *38*(7), 3668–3675. https://doi.org/10.1016/j.enpol.2010.02.043

Victorian Government. (2017). Microgrid Demonstration Initiative. https://www.energy.vic.gov.au/renewable-energy/microgrids

Victorian Government. (2019). Solar Energy Facilities: Design and Development Guideline (August 2019).

Victorian Government. (2020). Community Microgrid and Sustainable Energy Program. https://www.energy.vic.gov.au/renewable-energy/microgrids

Victorian Government. (2021a). Neighbourhood Battery Initiative. https://www.energy.vic.gov.au/new-energy-technologies

Victorian Government. (2021b). Solar for Community Housing. https://www.solar.vic.gov.au/solar-community-housing

Victorian Government. (2021c). Community Engagement and Benefit Sharing in Renewable Energy Development: A Guide for Renewable Energy Developers. Second Victorian Renewable Energy Target Auction (VRET2). https://www.energy.vic.gov.au/__data/assets/pdf_file/0036/536787/Community-Engagement-and-Benefit-Sharing-Guide.pdf

Victorian Government. (2021d). Development of Wind Energy Facilities in Victoria: Policy and Planning Guidelines (November 2021).

Vringer, K., & Carabain, C. L. (2020). Measuring the legitimacy of energy transition policy in The Netherlands. *Energy Policy*, *138*(January), 111229. https://doi.org/10.1016/j.enpol.2019.111229

Wassermann, S., Reeg, M., & Nienhaus, K. (2015). Current challenges of Germany's energy transition project and competing strategies of challengers and incumbents: The case of direct marketing of electricity from renewable energy sources. *Energy Policy*, *76*, 66–75. https://doi.org/10.1016/j.enpol.2014.10.013

Weber, M. (1978). *Economy and society: An outline of interpretive sociology. (G. Roth & C. Wittich, Eds.).* University of California Press.

Walker, G., & Devine-Wright, P. (2008). Community renewable energy: What should it mean? *Energy Policy*, *36*(2), 497–500. https://doi.org/10.1016/j.enpol.2007.10.019

Wiseman, J., & Wollersheim, L. (2021). *Building prosperous, just and resilient zero-carbon regions: international experience.* Melbourne.

Wolf, I., Fischer, A.-K., & Huttarsch, J.-H. (2021). Soziales Nachhaltigkeitsbarometer der Energie- und Verkehrswende 2021.

Wunderlich, C. (2012). Renews Spezial Akzeptanz und Buergerbeteiligung fuer Erneuerbare Energien, Berlin, 2012.

Wüstenhagen, R., Wolsink, M., & Bürer, M. J. (2007). Social acceptance of renewable energy innovation: An introduction to the concept. *Energy Policy*, *35*(5), 2683–2691. https://doi.org/10.1016/j.enpol.2006.12.001

Yildiz, Ö., Rommel, J., Debor, S., Holstenkamp, L., Mey, F., Müller, J. R., Radtke, J., & Rognli, J. (2015). Renewable energy cooperatives as gatekeepers or facilitators? Recent developments in Germany and a multidisciplinary research agenda. *Energy Research and Social Science*, *6*. https://doi.org/10.1016/j.erss.2014.12.001

Ziekow, J. (2018). Planungsrecht. In H. Lars, & J. Radtke (Eds.), *Handbuch der Energiewende und Partizipation (1st ed.)* (pp. 333–353). Springer VS.

Comparing local energy conflicts in NSW Australia: moving to climate generosity

Jonathan Paul Marshall

ABSTRACT

Australia does not have 'carbon democracy', but a 'carbon oligarchy,' in which the State defends the established and self-destructive relations of order, wealth and power built around fossil fuel energy. The article sketches interaction with carbon oligarchy in three NSW country towns of Lismore, Narrabri, and Bega, in relation to renewable energy. The aim is to uncover some problems of energy transition, and to suggest possible courses of action. In one town, a Local Council program to become self-supporting on renewables ran into problems of regulation and neoliberal economic values. In the second, transition was constrained by local fossil fuel supplies. In the third, some people embraced a form of 'climate generosity' which enabled support for transition against the requirements of carbon oligarchy. Climate generosity could usefully supplement Climate Justice, by circumventing demands for fairness and by providing an exemplar of action to help build community, support and meaning.

Introduction

Recent history implies the improbability of current modes of State or business organization having the capacity to prevent increasing climate and ecological crises. Obstacles abound. Burning fossil fuels is a major source of the greenhouse gases causing climate change, yet major struggles occurs over the transition from fossil fuel based energy to renewables. To elucidate this problem, this field-work-based study focuses on attempts at energy transition in three country shires in New South Wales (NSW), Australia – Lismore, Narrabri, and Bega – suggesting that some of these problems reside in 'carbon oligarchy' and its neoliberalism. It examines difficulties in the different pathways attempted, and suggests that localized groups of people may gain greater freedom to act through 'climate generosity'.

These country towns were chosen by accident, but contrast in their approaches to energy transition, facing different challenges in their responses to climate change and the implementing of renewable energy projects. In Lismore, a program for the Local Council to become self-supporting in renewables ran into problems of delay, cost, regulation, confused state policy, and maintenance of economic values. Established modes of functioning undermined transition. In Narrabri, transition seemed hindered by abundant, Federal Government supported, fossil fuels, together with a mining company's exploitation of historical fractures between town and country. Deferral resulted from the oligarchy maintaining power and profitability. In Bega, *some* local people embraced a generosity which presses against neoliberal economic 'common sense' and helped facilitate action. This latter response suggests that 'climate generosity' can supplement climate justice, help bypass the oligarchy, and promote a successful energy transition. Research in Lismore was done by myself and a PhD student Alana West, in Bega by myself, and work in Narrabri involved undergraduate students and other academics.

The world crisis

The delay in acting

Climate change has been recognized for many years. The role of fossil fuels in generating climate change, has been established even longer (Bell, 2021). Margaret Thatcher announced its significance in the late 80s (Thatcher, 1989). The UN Framework Convention on Climate Change was signed by 154 states in June 1993 and the Kyoto Protocol approved in December 1997.

Despite this recognition, the consumption of fossil fuels has increased. Using figures from the IEA, burning rose from 7,115,986 'kilotonnes of oil equivalent' (ktoe), in 1990, to 11,596,919 ktoe in 2018. During the same period, wind and solar increased from 36,571 ktoe to 286,377 ktoe in 2018. In total, fossil fuel consumption has increased by over 50% on 1990 figures and the world currently uses 46 times more fossil fuels than solar and wind together (IEA, 2020). A small decline occurred during 2020, due to Covid, but a large rebound was expected (IEA, 2021). Despite 25 years of global meetings and targets, the problems of emissions and ecological destruction have increased.

To avoid runaway climate change humanity needs to replace fossil fuels. This, in turn suggests the necessity (and problems) of a joint massive build of renewables *and* a massive phase out of fossil fuels. A recent news release from the UN states:

> current levels of climate ambition are very far from putting us on a pathway that will meet our Paris Agreement goals. (UN Climate, 2021a)

A report by Climate Action Tracker from September 2021 estimates emission levels will remain constant until 2030, rather than declining by half, which is necessary to remain under a 1.5 degrees increase:

> Australia, Brazil, Indonesia Mexico, New Zealand, Russia, Singapore, Switzerland and Viet Nam … have submitted the same or even less ambitious 2030 targets than they had put forward in 2015. (CAT, 2021)

Current international programs are not succeeding.

Australian paradox

The Australian Federal Government will not increase 2030 targets as recommended by COP26, and largely relies on imagined developments in technology to reach net-zero by 2050 (Australian Government, 2021). This lack of government action, in accordance with world failure, contrasts with apparent popular enthusiasm. Australian polls are consistently pro-renewables and pro-climate action, but do not always translate into electoral success. Polls may not predict what people will do, but they can show aspirations, and preferred trends. *The Guardian*'s 'Essential' poll in September 2020 found two-thirds of voters wanted the government to support renewables rather than gas plants. This popularity crossed 'all gender, age and voting demographics'. Even voters for the governing parties were 60% in favour (Murphy, 2020). A Lowy Institute Poll (2021) claimed 60% of Australians view climate change as a pressing problem while 91% supported government 'providing subsidies for the development of renewable energy technology'. A poll in the rural Bega region found 59% of the sample would 'be more likely to vote for a candidate who supported publicly-funded renewable energy projects to create secure, full-time jobs in the local area' (Hurst, 2020). This is common elsewhere. A Pew poll of people in 17 countries, found 72% of people fear that climate change will hurt them personally and 80% would make changes in their lives to beat climate change (Bell et al., 2021).

The popularity of renewables in Australia is also shown by rooftop solar installs, being 18% higher in 2020 despite Covid. By 'the end of 2020, an estimated 2.66 million [of about 10 million] Australian homes and businesses had a rooftop PV system' (AEC, 2021, p. 3). Action also seems popular with many local Councils. ClimateWorks and Monash University's Sustainability Unit, surveyed 57 local Councils and found 58% of the local governments assessed had a target to reach net zero operational emissions by 2050 and the cities of 'Sydney, Melbourne and Brisbane are already carbon neutral for their operational emissions' (ClimateWorks, 2020).

Clearly higher level State action in Australia and elsewhere has been inadequate and has not matched local enthusiasm or apparent desire. When organizational forms repeatedly fail to deliver their intended results, then that failure is part of their inherent dynamics, and an analysis of those dynamics is needed to move forward. Understanding of solutions will fail if theories about how humans organize, or become motivated, are inadequate. The study of failure is as important as success. This paper suggests that Carbon Oligarchy and neoliberalism, set up situations of failure, which can undo even spirited attempts at transition, and that climate generosity might form a way out from this restriction.

Carbon Oligarchy

Timothy Mitchell in his book *Carbon Democracy* (2013) argued that the organization of coal fired energy changed class relationships, because workers could relatively easily disrupt the mining, distribution, and burning of coal. This disruptive presence forced dominant social groups to negotiate with workers *in general*, leading to extended democracy and the beginnings of a relatively protective and helpful State. When the dominant groups turned to oil, then by design or accident, they organized oil in ways which suppressed the workers' power found in the coal industry, while keeping control over energy supply and use: 'politicians saw the control of oil overseas as a means of weakening democratic forces at home' (Mitchell, 2013, p. 36). Pipelines removed potential blocking points, and massed worker participation in oil distribution was reduced, so workers had less power (Mitchell, 2013). States depended on fossil fuels to promote the growth of the modern economy and military. This magnified the power of those providing the fuels.

It can be suggested that representative democracy has now been replaced by a Carbon Oligarchy (Mitchell, 2013, pp. 8–9, 122, 143), which seems to have transferred power back to the owners of coal, and forward to gas. In Australia, it is widely believed fossil fuel and mining companies sponsored the overthrow of at least one prime minister, gives post-politics jobs to politicians and takes positions in government (Wilkinson, 2020; Baer, 2022, pp. 59–60). The current Prime Minister's Chief Of Staff (John Kunkel) was 'Chief Adviser, Government Relations' for Rio Tinto and 'Deputy Chief Executive Officer of the Minerals Council of Australia'. The head of the National COVID-19 Commission Advisory Board (Neville Power), which recommended a 'gas led recovery', was the CEO of Fortescue Metals and Deputy Chair at Strike Energy, a gas company.

There is no guarantee renewables will change power relations (Baer, 2022: xi). Tronti (1965) points to the 'history of the successive attempts of the capitalist class to emancipate itself from the working class', while some supplementary technologies to fossil fuels, such as carbon capture and storage, could function as attempts to emancipate capitalists from the 'limits of nature'. Mitchell also elaborates how carbon oligarchy, commercial power and neoliberal theory reinforce each other, creating modes of common sense and procedural rationality (2013:, p. 141ff. 197ff. 224). Walker (2020) further shows how neoliberal ideologies were used to prevent environmental action from impinging on both corporate 'liberty' and the hegemony of fossil fuels, by forcibly decrying governmental intervention in markets, attacking environmental regulation as 'green tape' and promoting 'market based' environmentalism. Dario Kenner (2019) argues that political and economic elites, are a 'polluter elite'. Wealth, power and 'freedom' are tied to pollution through: consumption, travel, luxury imports, and energy wastage; investing in polluting and destructive industries; and protection of the freedom to pollute through 'State capture.' He says:

> Given the dominant political power of the oil and gas polluter elite the low-carbon transition will only happen on the large scale and at the rapid speed that is needed if they are weakened. (2019:, p. 10)

It seems the 'Modern World' has been built on destructive but cheap fossil fuel based energy, and has generated a social ecology of technology, regulation, and procedures to support that technology, which hinders attempts to escape it.

The presence of a carbon oligarchy and its networks seems obvious in Australia (Wilkinson, 2020; Baer, 2022, p. 53, 56). The Coalition-led Australian Federal Government has generally supported fossil fuels. BloombergNEF (2021, p. 14), states that Australia increased financial support for fossil fuels by 48% between 2015 and 2019. John Hewson, ex-leader of the Australian Liberal Party (the main part of the current Federal Government) said, while visiting Bega, one of the towns in this study:

> In the thirty years that I have been involved in this issue, we still don't have a national energy policy, a climate action strategy, or an emissions reduction strategy. (q. Campbell, 2019)

In Australia, the major parties have supported renewables, fossil fuel power stations and fossil fuel exports which do not count in Australia's official emissions (Baer, 2022). The current Government supports a taxpayer funded 'gas led recovery', with new gas fields, pipelines, and power (Morrison, 2020). Australia had a carbon price, under the previous Labor government, which was removed in 2014 after the Coalition took power, and replaced by a series of unclear and even hidden policies, which made planning difficult (Wilkinson, 2020, pp. 189–202, 227). Participants in this fieldwork commonly complained about the difficulty of steering through regulations or discovering what aids to progress were available. Uncertainty and confusion seemed normal. This may be lessening at the NSW state level, with the government promoting Renewable Energy Zones, but new coal and gas for export seem more openly encouraged than community energy.

Three NSW country towns

Comparing towns of a similar size under a similar regime allows the possibility of observing what strategies seem the most and least successful; with the potential of having implications elsewhere. Town choices were reinforced by their differences. In Lismore the Council was an early mover, with a renewable energy target for Council operations which looked achievable. Narrabri was chosen because of the heavy presence of fossil fuels, and Bega arose because of my attending a climate event organized by Clean Energy for Eternity in Tathra, a small beach town in the area, and finding their 'climate generosity' inspiring and apparently effective. Data collection has been conducted primarily by interview, reading local papers and official reports, hanging out with participants, and attending events in the areas, to get some sense of how the social dynamics and conversations work. These are not stories of finished triumph, they are stories of struggle and possible failure, occurring within the political ecology of the Carbon Oligarchy. This political ecology manifests primarily as a technology of regulations and rules built up around an understanding of correct procedure, economy, values, and common sense, which hinders popular action and success in changing that system. This establishment is challenged by climate generosity. How this will work in the long-term is unclear, but it seems to have considerable benefits.

Lismore

Lismore, is a town and a shire in the Northern Rivers area of New South Wales. It has a history of activist campaigns, most famously preventing logging in the Terania Creek rainforest (Bible, 2018), together with the Bentley Blockade, and Gas Free northern Rivers Campaigns (Kia & Ricketts, 2018). Success in energy transition would be expected and began early at both household and council levels. In 2014, Lismore was reported to have 'a higher percentage of households ... using rooftop solar and/or hot water systems than anywhere else in Australia' (Mortimer & Park, 2014). Currently 42.4% of households have solar panels (APVI, 2021). However, the program for Council to move to 100% renewables has, at the moment, lost momentum through problems of cost, regulation, hidden state policy and the maintenance of economic values. Lismore shows the problems of innovating by rule following.

In 2005 the Council embraced a goal of 20% emissions reduction in 10 years, starting with energy efficiency and attempts to reduce expenditure on Council electricity bills despite large increases in electricity costs. In 2009 the first solar panels were installed on Council buildings. In 2012 they began 'Imagine Lismore' which involved consultations with randomly selected citizens, finding massive local support for renewables, and they announced a 'challenge of becoming 100% self-sufficient on electricity from renewable resources by 2023' (Lismore Council, n.d., p. 30), becoming 'the first regional Council to commit to a renewable energy goal' (Interview 2, 21 March 2018).

This led to the 2014 Renewable Energy Master Plan (Lismore Council, 2014) which formalized the self-sufficiency target.

> [Council] wanted to identify two sites that ran 7 days a week, so they had 7 day use and ... wanted to develop a community funding partnership model. (Interview 2, 21 March 2018)

They worked on the funding model with Farming the Sun, a 'registered charity organisation [which had] ... already done a number of [community projects]' (ibid), and decided to borrow from the community, paying back with interest.

Borrowing money from the community for Council funds had never been done in NSW, and was difficult to organize. The solution involved Farming the Sun raising the money through a share offer and then issuing a loan to Council for seven years. When the loan is paid off, Council owns the farm.

> Council pays ... 5.5 per cent interest and the investors get a franked dividend every 6–12 months ... an attractive investment due to low bank interest rates. (Interview 2, 21 March 2018)

Reportedly some Councillors argued for a cheaper loan from a bank. Council workers explained the Council wanted community involvement and that 'it's more than just the bottom line, it's meeting our partnering strategy, taking us towards our goal to be self-sufficient and our community's vision' (Interview 2, 21 March 2018). The Council's own guidelines reinforced the policy and helped get the initiative passed. However, focusing on ensuring local investors got a good rate of return restricted what they could do.

Another problem was lack of trust in the stability of government policy:

> there was this political uncertainty around the Renewable Energy Target. If your system is under 100 kilowatts you could claim the Renewable Energy Certificates ... rebates up-front. [Above] 100[kw], you had to claim them on an annual basis. (Interview 2, 21 March 2018)

Council were uncertain how long the government would maintain the certification system. If it ended before the seven years was up then they could not guarantee a fixed rate of return. This uncertainty limited available options so they decided on a system under 100 kw to guarantee Certification. Consequently the project was smaller than desired. Regulatory problems continued. Due to Australian Securities and Investment Commission requirements, shareholdings had to be 'limited to only 20 shareholders per site because it was a private share offer'. As $180,000 was needed for each site, this meant the shares cost $9,000 each.

Lismore is not a wealthy area, and 'we were pretty nervous about whether people would invest' (Interview 2, 21 March 2018). However, the community pledge was filled in 'a matter of weeks' (Hunniset & Powell, 2017) and in October 2015 Lismore City Council approved borrowing $360,000 in total. Council could have raised money for an extra solar farm given citizen interest in the loan (Vorath, 2016), but chose not to. The farms were placed on the roof of the swimming centre and in ponds at the sewage farm.

Regulations also caused difficulties with using solar on one site to power another site, and hence siting of the projects:

> we had a perfectly sized roof at the waste facility and it's only 400 metres away ... But because they're on two different meters, it just wasn't possible ... the electricity would have cost us more because they would have charged us for the network fees twice. (Interview 2, 21 March 2018)

Sites were limited to those able to use all the energy produced. Later plans to electrify the central shopping centre with solar panels, while upset by flooding in 2017, were scrapped due to similar cable and regulatory problems. One member of Council management complained the regulations made it:

> too hard ... all I can see is we will be embroiled in some oxygen energy sucking machine that will take us away from every other strategic and operational thing we have to do. (Interview 3, 24 June 2019)

Attempting to fulfil regulations sapped initiative and limited results.

The sewage farm plant was also delayed by unfamiliarity and regulation. Most companies contacted did not use floating panels, and some wanted millions of dollars 'because they were ... having to manufacture something they had never made before' (Interview 2, 21 March 2018). Council found a company in South Australia which had become a distributor for a French floating product not yet installed in Australia. However, 'when we got the designs ... [they] didn't comply with Australian Standards'. Consequently the product had to be redesigned, with new safety walkways: 'it ended up taking 6–12 months because ... they had to make new moulds and stuff' (Interview 2, 21 March 2018).

Construction began in October 2017, with the expectation of expanding the farm to power the whole sewage site (Harper, 2017), and begin an ongoing expansion of renewables. The Mayor announced they were 'looking at a mid-size solar plant somewhere else around the LGA' (Anon., 2017).

A local newspaper reported the expected benefits; financial and communal, through becoming an exemplar, and through expected future progress:

> Sharyn Hunnisett [says] ... "the money we spend remains within the Lismore community", [and] "There's been national interest in the floating solar farm and we have been invited to share our success story .... this is the first step and ... will definitely be expanded". (Paterson, 2017)

Lismore Council won awards for their farms and process.

Those involved in the project suggested it took 4–5 years of background work because of the regulatory and political complications and uncertainties, and the need to guarantee investment returns. This took energy away from other projects. Another obstacle emerged when the General Manager who instigated the renewable program resigned. The new GM discovered a budget deficit (which appears unconnected to energy), and changed priorities. Some people allege this change led to a crisis of communication, where uncertainty dominated and delayed the renewable program for over a year:

> it's mad not to do it, it saves money ... but I don't know that we will be doing anything outside the square in the next 12 months. (Interview 4, 26 February 2019 conducted by Alana West)

However, budgeting rules were changed with the outcome that 'they were either going to defer projects or cancel them altogether' (Interview 5, 18 June 2019).

In 2020 the environmental officer's job, which included planning the renewable energy transition, was scrapped. This was apparently not formally discussed in Council, nor reported locally in advance, indicating the Council's work was not embedded in the community. While Council did plenty of consulting, they did not help the community organize themselves, nor am I aware of close working relations between Council and pro-renewables groups.

At the time of writing it appears that progress has halted with the Council coming against neoliberal framings of debt, and what is defined as economically essential. Council could not escape the neoliberal common sense of managing finances by cutting spending, assets and services, and so avoided looking at future climate problems. While there is a possible further change of direction, after Council elections in 2021, it seems that Lismore tried to conduct transition by the rules in an orderly manner, but were caught by the Carbon Oligopolies' regulations, confusion and crisis.

Narrabri

Narrabri is an inland town and shire in NSW which, since colonization, has grown through farming. With little rain and one main river, most farms survive via bore water. One of its major agricultural industries is cotton farming, which requires considerable water. Narrabri can be said to suffer from the resources curse, as it sits upon plentiful supplies of coal and coal-seam gas beneath the water table. This situates it as a region of interest for the Carbon Oligarchy, which seems determined to gain access to local fossil fuel resources before they are stopped by slowing demand or legislation. This rush to exploit helps to maintain the fossil fuel lock-in in possible times of change, creating the so-called green paradox of increased pollution due to the possibility of green policies (Sinn, 2012). Narrabri has had a history of popular action against gas fields, largely because of feared damage to water supplies. These features strongly affect discourses about renewable energy.

People in the town often seemed to claim Narrabri was in crisis: social, economic and ecological. However, 'sustainability' seemed framed in terms of maintaining, or even increasing, population, rather than maintaining ecology. Young people leaving the area was a major concern (cf Coffey et al., 2018). One influential person described this as 'our most precious, valued export' (Interview 6, September 2018, research students). High school students we talked to, wanted to leave, displaying little knowledge that 'adults' and Council were concerned about them leaving (Forum 1, September 2018, research students).

Solutions to 'sustainability' problems focused on increasing employment and monetary prosperity. For many, it did not matter where the jobs came from, with the obvious solution being more mining for fossil fuels, although this provoked large scale community dispute. We were told of long-term friendships which had broken due to the conflict. People reported splits in sports clubs between those who would accept the 'community funds' provided by the companies and those who would not. Others reported repeated abuse. One group alleged 'the industry have ... come in [and] divided the community' (Coonabarabran Discussion Group, 2018). This fracture seemed to be experienced as deeply painful by many community members, although some said the stress was diminishing.

In the region there was little to no contest between large-scale renewable energy and fossil fuels. Renewables were rarely held to promise long-term employment despite an Institute for Sustainable Futures report supporting an ambitious renewables scenario taking up significantly less land than gas exploration while creating '3,600 jobs and produc[ing] 3,900 MW of power' (ISF, 2018). Fossil fuel companies had a presence in, and a history with, the town, and had donated plenty of money to town activities. In comparison, renewables companies seemed to have little regular presence in town at least until 2019, just before COVID, when a local organization developed plans to build a community owned virtual power plant. During the fieldwork period people interviewed seemed only vaguely aware of, or uninterested in, plans for large scale solar farms in the region (cf. Narrabri Council, 2020a, p. 41). It was unclear what would proceed and some people suggested these plans were just ambit claims to drive off competition. Solar farms seemed a distant, almost alienated, form of energy, and conservationists opposed to coal and gas could also oppose wind farms. Unlike Lismore or Bega, Council seemed largely uninterested in building any renewables themselves, although the *Local Strategic Planning Statement* identified community 'support for solar or wind renewable energy developments' while pointing to 'the need for immediate upgrade to interconnecting grids' for commercial renewable farms. It also noted the common problem of 'lack of certainty around State and Federal government energy policy' (ibid, 41). Council's Annual report claimed that 'plans for installing energy efficient technologies and innovations ... i.e. solar panels, LED lighting, air-conditioning' were put on hold 'due to budget restraints' (Narrabri Council, 2020b, p. 25), again showing neoliberal regulation in action.

Those interviewed were aware of the possible dangers from gas, and the devastating experience of other communities, but local Councillors (for example) thought they were more prepared than others had been. Business people we talked to seemed to accept the gas company's promises of no contamination, even though there had already been some local accidents, and it was impossible to guarantee seals on drilling sites, and no risk to the water table, forever. The Carbon Oligarchy's promises seemed largely unquestioned outside of activist groups. Opponents of the drilling frequently mentioned that it was unclear whether gas jobs would go to locals rather than fly-in-fly-out workers, or temporary workers (who would get counted as locals). As another report argued:

> the recent expansion of coal mining and CSG exploration in the Shire have engendered considerable conflict between those who see the extractive industries facilitating economic growth and diversity

> and those who see these activities as a threat to the core agricultural functions upon which Narrabri Shire is founded. (Askew & Askland, 2016, p. 17)

These contending positions seem fuelled by an historical rivalry between town and country. Many people claimed the status of farmers had declined, and they were no longer the most prosperous section of the community, a situation magnified by prolonged drought. Farmers' relative decline in income, plus new technology, meant they employed fewer townspeople as casual labourers and spent less in the town. This weakened ties, dependencies, and mutual concern, helping the gas company promote their plans. Farmers were closer to the areas being affected by mining and thus more reluctant to support it. Those nearer the coal mines were affected by dust, and worried about river pollution. Those near the gas seemed more opposed than the townsfolk because of bore water issues. However, declining income provoked fears that some farmers would rent out their land for exploratory gas wells, potentially risking their, and others, bore water (cf Hunjan, 2018).

However, the issue seemed out of local hands, as the Federal Environment Minister approved the gas fields (Santos, 2020b), encouraged by a deal between the Federal and NSW governments under the Federal Government's 'gas led recovery' and National Gas Infrastructure Plans (Santos, 2020a). Narrabri's fate seems largely dominated by those seeking security through the Carbon Oligopoly, resource extraction, anticipated 'jobs' and neoliberal common sense, although there is continuing opposition to the path being taken. Whatever the result, climate change, water risk, drought, and Government policy have all made decisions extraordinarily difficult for local people. It is not an enviable position to be in. Neither of the other two areas currently face such severe problems.

Bega

Bega is a town and Shire in Southern NSW, primarily renowned for its dairy farms, lush fields and bush. Like Lismore it has a mixture of established agricultural families and people who moved in for the quiet life. However, the area faces long-term drought, and has suffered ferocious bushfires almost certainly intensified by climate change. The fires severely affected some small towns and burnt down many houses. Bega is the source of some interesting conflicts and developments over the energy transition.

I came to Bega excited by something I had not seen before. Local people, particularly in the local town of Tathra, were embracing a form of generosity undercutting the neoliberal economic common sense of Carbon Oligarchy. The community organization, Clean Energy for Eternity (CEFE), proceeds by gifting solar panels for the rooftops of community buildings. They ask local people to donate money or panels, and have 'installed renewable energy on every community building in Tathra' (CEFE, 2019). They use payback from the gifts (from saved electricity costs) to purchase more gifts for the community. For example, they:

> developed the 30 KW "Imagine" Solar Farm, in partnership with the Bega Valley Shire Council [by gifting panels to Council], saving Council dollars .... We invest these savings into renewable energy installations on community buildings, in partnership with local organisations. (CEFE, 2019)

The farm was expected to supply '50 per cent of the power needs of the sewage treatment plant, which currently costs about $30,000 a year to run' (Bega Anon, 2015), and potently demonstrated the feasibility of solar to Councillors. CEFE further organizes bulk buys of solar panels with low interest finance for purchasers, again ploughing any profit back into more panels gifted to public buildings.

This approach bypassed the rules which hurt Lismore and the supposed economic realities restraining Narrabri, setting in place 'climate generosity.' As well as deploying generosity, CEFE focused on building community and engagement through 'happenings', meetings, events, prominent speakers, entertainments, human signs, and so on. This helped build good feeling, usually with a positive objective, such as putting panels on a surf club roof, or asking 'the Bega Valley community to commit to a 100% renewable energy target by the year 2030' and issuing a declaration when the motion was passed (CEFE, 2019). These practices both reinforce and demonstrate community involvement, which then leads to Council involvement. Public gifting builds relationships, creates exemplars, builds obligation, and can instigate public action. In theory everyone who wants to can participate. CEFE is hard to object to as they are giving to the community, or to local Council as a symbol of community, and perhaps for the sake of future generations, to show that people want something done about climate change. Furthermore, the gift can occur because the cheapness and modularity of solar energy makes it easy for people to participate as money becomes available.

Bega Council is caught in severe conflict over climate change and renewables. One Councillor said:

> in terms of Council, I can go on until the cows come home about the argy-bargy there (Interview 7, 20 March 2019)

Several Councillors seemed to be dedicated opponents to renewables, however, CEFE's procedure demonstrates committed community support and involvement; people turn up to their events and talks and put in money with visible results. Those Councillors favouring energy transformation know they have significant community support, which helps them stand against 'neoliberal' opposition. They have been able to declare a climate emergency (which directs Council to check how Council proposals interact with climate change in the region) and Bega's climate resilience action plan was passed relatively easily; neither seemed possible at Narrabri. Bega Council still has to act according to the regulations, but it keeps encountering community input and support for action. Rather than a few isolated people joining in a loan, many local people are putting money and effort into visible projects. CEFE and the generosity framework's importance was emphasized when several people informed me that attempts to establish a new Bega Council-community solar farm, as in Lismore, foundered after similar problems with regulations and attempts to guarantee investment returns. This severely constrained possibilities and the venture failed.

CEFE's actions attracted attention from outside. The, then, Leader of the Federal Opposition, Bill Shorten, announced a major renewable policy initiative in the region during the 2019 election campaign:

> What we're going to do is allocate $10 million, to provide Tathra and surrounding communities with the opportunity to embrace the clean energy revolution. (Shorten, 2019)

If the election results had differed, these projects could have made a country wide difference, which could have been attributed to small actions by CEFE.

While CEFE and its allies move from a position of strength, certainty and generosity, there are potential problems. Massive fires over the 2019–2020 summer, followed by the Covid-19 outbreak, took energy away from the process. CEFE could not organize public events due to lockdown, although they did gift panels to some rural fire sheds, and attempted to get the Rural Fire Service to collaborate on putting panels on all rural fire sheds, but failed (Interview 8, 26 November 2020). CEFE has only a few dedicated older committee members. While their processes magnify the impact of these members, CEFE is vulnerable to generational change. Winning government support is still vital to transition, but CEFE's method helps generate visible local movement and support, while establishing community energy. The question is whether this can apply on a larger scale, beyond face-to-face encounters at the local level.

'Climate justice' and 'Climate generosity'

Apart from simple survival, one of the main sources of opposition to the Carbon Oligarchy has come from Climate Justice. The points made are important. For example, the main victims of climate change are:

> the poorest and most marginalised in both the Global North and the Global South, who frequently lack access to the economic, social and political structures necessary to ensure that their views are recognised, their interests represented and their needs addressed. (Jafry et al., 2019)

The Mary Robinson Foundation (n.d.) says climate justice is about 'sharing the burdens and benefits of climate change and its impacts equitably and fairly' while the UN Environmental Program (n.d.) points out 'Climate justice is a term used for framing global warming as an ethical and political issue'. However, this framing *may* hinder action and is relatively easily incorporated into the neoliberal profit framework.

Not only is justice *not* a universally resolved concept granting clarity of action, but many refusals to act on the ecological crises use 'justice' or 'fairness' arguments. For example, people claim Australia is responsible for just over 1% of global emissions and therefore has negligible effect on the world, suggesting it is neither 'just' nor fair to ask Australians to 'risk the economy' or neoliberal power (McDonald, 2019). The Prime Minister (Morrison, 2019) made such an argument at the UN. The reality is Australia has much less than 1% of the world's population, has a very high emissions rate per head of population, and exports massive amounts of fossil fuels which are not counted as part of Australia's emissions, but the argument can still stand: it's not fair or 'just' to act alone.

'Developing' countries can argue that while their increasing emissions and development are likely to magnify the crises, it is 'just' for them to continue. The West had years of unconstrained growth and developing nations should be able to catch up. From this perspective, attempts to stop these countries from polluting are attempts to stop poverty reduction, or attempts by the West to maintain domination. Allowing pollution is a matter of justice. Australia politicians use this argument to justify selling coal to India and China: 'we actually are quite passionate about the economic success of countries in our region' (Morrison, 2021). Claims that the West should cut back to zero *first*, can be met with the response that developed nations should not cut back either, because where is the justice or fairness in crippling their economy and hurting their people, to allow others to pollute massively? In both cases the Carbon Oligarchy's move of putting 'the economy' over ecology, and implying development requires fossil fuels is naturalized as 'just'. It may take forever for a purely 'equitable' and 'just' situation to arise. Consequently, climate justice can provide a rhetoric for keeping things as they are, and unintentionally protecting the Carbon Oligarchy.

Climate Justice was not a commonly discussed mode of activating people in these towns, as they could wait for others to cut emissions first. In Narrabri, it could be argued the town could only survive through mining. Some people interviewed in Bega and Lismore were worried about poorer people not being able to afford renewables, and having to use expensive, polluting, energy, but there is little evidence locals were doing much to remedy this, now.

'Climate Generosity', on the other hand, proposes that we give to others without waiting for fairness. 'We' act to get things done and because we can act now. No one needs to be punished or forced. Generosity does not seek blame, nor worry the economy will be destroyed, but asserts the right to action (and agency) by people and governments, at this imperfect moment. Even the relatively poor and powerless can act and gain respect and agency (Rudiak-Gould, 2014).

Gifting is a common human behaviour, which ignores neoliberal economics. In many societies people give generously to others to gain respect, or persuade others to give in return, and to build alliances, active trust or reduce the noise of communication (Klapwijk & Van Lange, 2009; Przepiorka & Liebe, 2016). Even small acts of generosity can build networks and alliances (Whitham, 2021). 'Gifting' lies both outside and within capitalism, and although classically a gift may expect return, equal reciprocity is not a requirement, and relationships do not end when the gift is given (Cheal 1988, p. 126). As Blau suggested exchange is caught between pure calculation and pure generosity; neither of which may exist unmixed (1964, p. 112). Gifting can motivate other people to act, shaming inactivity. Through seeing others acting, people can come to think they can act themselves, without waiting for others. Generosity is usually defined as 'good', hence people may emulate those being generous in a positive feedback loop, adding to the climate gifts becoming available, and helping solve the problem. It demonstrates that the givers are motivated to act. As such they can become an electoral force needing to be acknowledged. Further, they can display support for change, and can be relied upon by those active in pursuing the same ideals. Generosity also seems to make givers feel good, builds meaning and connection, while undermining the acceptance of neoliberal transactionalism and claims profit is fundamental to human psychology (Smith & Davidson, 2014).

With justice you have to wait for a framework for justice to occur, but with generosity people can act immediately. They don't need State action to begin with. Generosity can be a daily behaviour. If a person gives solar panels or contributes to wind turbines, nobody loses, everybody benefits including themselves; eventually it could seem odd not to support giving. Generosity slashes through indecision, or the restrictions of profit seeking behaviour; you are doing this because you want to and it's rewarding. It could cross political barriers, as humans tend to recognize generosity in action. Generosity does not pretend we are only competitive, but people can compete in their generosity, should they choose to. The many ways of being generous allows a multiplicity of ways to approach multiple eco-crises. While co-operation and alliance can be built through generosity, whether this can apply at the international level is uncertain, but the current framework is not functional, and processes of generosity could break the deadlock.

Generosity is harder at the international level because of interactional histories between States, but it is possible if States give with care, and without demands (Barnett & Land, 2007). However it is easily possible for people to take this motive up in their own communities, build community energy without profit requirements, and through taking advice from those organizations already engaged in Climate Generosity.

Conclusion

Looking at these three country areas as exemplary forms of action common in the world, the following issues and blockages become clear.

Lismore

The Council attempted to follow the rules and build in profit for investors, while finding obstruction in those rules and normal priorities, losing momentum. The rules grew up under Carbon

Oligarchy neoliberalism and take that oligarchy's ways of action as fundamental. Regulations support the old order and old economy and led to reassertions of the old ways when Council faced the unrelated problem of debt. Despite wanting community involvement, Council began a top-down process with little active community participation. Council values may have been in the right place but being primarily rule-based and investor-oriented they slowed and hindered progress significantly.

Narrabri

Many locals ignored the crisis promoted by fossil fuels, both direct threats of damage to water supplies and the indirect threats of compounding consequences of climate change. At least one powerful faction favoured attempts to solve the severe problems of employment by taking advantage of plentiful fossil fuels and the 'generosity' of the Carbon Oligarchy, reinforced by State and Federal Government support for fossil fuel companies and gas production. The proposed 'solution' to the employment problem created social splits, probably undermining the ability of the town to pull together to address problems with water supplies and ecological damage. Future damage can arise from present security. The strategy takes the neoliberal economy and the Carbon Oligarchy as a given, rather than something to change. The values being professed, and the actions encouraged by the presence of fossil fuel resources are in conflict with plausible predictions of future conditions. A new community energy movement could alter this, as they are focused on sharing rather than profit, but it is too early to say.

Bega

This region has middle class radicals who largely ignore the rules, and bend established practices, exercising influence out of all proportion to their numbers. They take an existing interest and motivate people further. The movement is largely bottom-up and free of demands for monetary profit, but depends on good local leaders who appeal to people to act beyond themselves in service to a cause, while benefitting the community and gaining local status for showing they care. The strategy builds a sense of community and engagement through meetings and events, thus making social meaning. It provides an obvious base for activist Councillors and Council workers, and circumvents or avoids neoliberal economic conventions in a practical way. It gifts the future and states clearly 'If our politicians refuse to show leadership, then we will do it ourselves' (Interview 9). Momentum is generally kept up, and provides support for sympathetic action despite the hostility of some Councillors.

Perhaps large states and businesses cannot save the world precisely because they are so big and so ingrained with the Carbon Oligarchy's established habits, power relations and rules. To the contrary, generosity not only promotes action, but undoes the common-sense of neoliberal economics which values corporate profit over survival and makes profit-taking fundamental to human life. It also undoes the Climate Justice bind of simply making a normative appeal. The model of generosity on principle can be scaled-up, beyond the local. There is no reason, other than support for the Carbon Oligarchy, that nations likewise cannot simultaneously cut back their own emissions and give renewables to other nations, rather than encourage them to buy coal or gas. Generosity may be seen as a mode of refusal of capitalism which gives direction, and shows how co-operation can be built.

Acknowledgements

Thanks is also given to all those who participated in interviews, and helped this project. Particular thanks to the reviewers and editors. UTS Human Research Ethics Committee Approval Number: ETH17-1177.

Disclosure statement

No potential conflict of interest was reported by the author(s).

Funding

This research was sponsored by an Australian Research Council Future Fellowship [Grant FT160100301]. 'Society and climate change: A social analysis of disruptive technology'. The views expressed may not be those of the ARC.

ORCID

Jonathan Paul Marshall http://orcid.org/0000-0002-3914-7697

References

AEC. (2021). Solar Report January 2021. Australian Energy Council. https://www.energycouncil.com.au/media/jv4blk2l/final-pdf-australian-energy-council-solar-report_-jan-2021.pdf

Anon. (2017, November 17). Energy poured into opening solar farm. *The Northern Star*, p. 3. [Via Proquest].

APVI. (2021). Mapping Australian Photovoltaic installations. https://pv-map.apvi.org.au/historical#10/-28.7564/153.2066

Askew, M., & Askland, H. (2016). *Local attitudes to changing land use – Narrabri Shire: Full report*. NSW Department of Primary Industries and the University of Newcastle. https://www.newcastle.edu.au/research/centre/csrrf/publications/local-attitudes-to-changing-land-use-narrabri-shire

Australian Government. (2021). *Australia's whole-of-economy long-term emissions reduction plan*. Australian Government Department of Industry, Science, Energy and Resources.

Baer, H. (2022). *Climate change and capitalism in Australia: An eco-socialist vision for the future*. Routledge.

Barnett, C., & Land, D. (2007). Geographies of generosity: Beyond the 'moral turn. *Geoforum*, *38*(6), 1065–1075. https://doi.org/10.1016/j.geoforum.2007.02.006

Bega Anon. (2015, March 19). Community solar farm in Tathra spells IMAGINE. *Bega District News*. https://www.begadistrictnews.com.au/story/2955674/tathra-community-solar-farm-an-australian-first/

Bell, A. (2021). *Our biggest experiment: A history of the climate crisis*. Bloomsbury.

Bell, J., Poushter, J., Fagan, M., & Huang, C. (2021, September 14). In response to climate change, citizens in advanced economies are willing to alter how they live and work. https://www.pewresearch.org/global/2021/09/14/in-response-to-climate-change-citizens-in-advanced-economies-are-willing-to-alter-how-they-live-and-work/

Bible, V. (2018). *Terania Creek and the forging of modern environmental activism*. Palgrave Macmillan.

Blau, P. M. (1964). *Exchange and power in social life*. John Wiley.

BloombergNEF. (2021). *Climate Policy Factbook: Three priority areas for climate action*. https://www.un.org/sites/un2.un.org/files/bp_bnef_climate_policy_factbook_071921_final35.pdf

Campbell, I. (2019, February 20). Tathra starts push for 100% renewable energy in Bega Valley. *About Regional*. https://aboutregional.com.au/tathra-starts-push-for-100-renewable-energy-in-bega-valley/

CAT. (2021). Global update: Climate target updates slow as science demands action. Climate Action tracker. https://climateactiontracker.org/publications/global-update-september-2021

CEFE. (2019, January 24). Subject: Position Statement to request Government support to help meet community targets for renewable energy for Tathra and Bega Valley. Clean Energy for Eternity. https://cleanenergyforeternity.net.au/files/2019/03/PosStat20190124.pdf

Cheal, D. (1988). *The gift economy*. Routledge.

ClimateWorks. (2020, March 2). Net Zero Momentum Tracker: Local Government Sector. ClimateWorks. https://www.climateworksaustralia.org/wp-content/uploads/2020/01/ClimateWorks_NZMT_Local-Government-Report_Jan20.pdf

Coffey, J., Threadgold, S., Farrugia, D., Sherval, M., Hanley, J., Askew, M., & Askland, H. (2018). 'If you lose your youth, you lose your heart and your future': Affective figures of youth in community tensions surrounding a proposed coal seam gas project. *Sociologia Ruralis*, *58*(3), 665–683. https://doi.org/10.1111/soru.12204

Harper, D. (2017, November 16). Australia's first floating solar farm soon to be buoyant in Lismore. *ABC News*. https://www.abc.net.au/news/2017-11-16/australias-first-floating-solar-farm-will-be-buoyant-in-lismore/9157878

Hunjan, R. (2018, May 28). When coal seam gas comes knocking. *Insight*. SBS. https://www.sbs.com.au/news/feature/when-coal-seam-gas-comes-knocking

Hunniset, S., & Powell, G. (2017, February 27–28). 'Stories of Success' Lismore Community Solar. *C4CE*. http://c4ce.net.au/congress/wp-content/uploads/2017/02/Stories-of-success-Lismore-Community-Solar.pdf

Hurst, D. (2020, May 25). Climate change critical issue in Eden-Monaro by-election as six in 10 voters say more action needed by government. *The Guardian*. https://www.theguardian.com/australia-news/2020/may/25/climate-change-critical-issue-in-eden-monaro-byelection-as-6-in-10-voters-say-more-action-needed-by-government

IEA. (2020). World Energy Balances. https://www.iea.org/data-and-statistics/data-tables

IEA. (2021, October 28). Despite some increases in clean energy investment, world is in midst of 'uneven and unsustainable economic recovery' – With emissions set for 2nd largest rebound in history. IEA Website. https://www.iea.org/news/despite-some-increases-in-clean-energy-investment-world-is-in-midst-of-uneven-and-unsustainable-economic-recovery-with-emissions-set-for-2nd-largest-rebound-in-history

ISF. (2018, May 16). Renewable energy could power the future of Narrabri. *Institute of Sustainable Futures, University of Technology Sydney*. https://www.uts.edu.au/research-and-teaching/our-research/institute-sustainable-futures/news/renewable-energy-could-power-future-narrabri

Jafry, T., Mikulewicz, M., & Helwig, K. (2019). Introduction: Justice in the era of climate change. In T. Jafry (Ed.), *Routledge handbook of climate justice* (pp. 1–9). Routledge.

Kenner, D. (2019). *Carbon inequality*. Routledge.

Kia, A., & Ricketts, A. (2018). Enabling emergence: The Bentley Blockage and the struggle for a gasfield free Northern Rivers. *Southern Cross University Law Review*, *19*, 51–75.

Klapwijk, A., & Van Lange, P. A. M. (2009). Promoting cooperation and trust in 'noisy' situations: The power of generosity.. *Journal of Personality and Social Psychology*, *96*(1), 83–103. https://doi.org/10.1037/a0012823

Lismore Council. (2014). *2014 renewable energy master plan*. Sustainable Business Consulting Pty Ltd. https://www.lismore.nsw.gov.au/page.asp?f=RES-PCH-64-58-30

Lismore Council. (n.d.). *Ten years of greenhouse gas emissions: Monitoring and reporting 2002 to 2012*. https://www.lismore.nsw.gov.au/page.asp?f=RES-HUQ-71-55-36

Lowy Institute. (2021). Climate poll 2021. https://www.lowyinstitute.org/publications/climatepoll-2021/

Mary Robinson Foundation. (n.d.). Principles of Climate Justice. https://www.mrfcj.org/principles-of-climate-justice/

McDonald, M. (2019, June 18). How to answer the argument that Australia's emissions are too small to make a difference. *The Conversation*. https://theconversation.com/how-to-answer-the-argument-that-australias-emissions-are-too-small-to-make-a-difference-118825

Mitchell, T. (2013). *Carbon democracy: Political power in the age of oil*. Verso.

Morrison, S. (2019, September 25). National Statement to the UN General Assembly. Prime Minister of Australia's Office. https://www.pm.gov.au/media/national-statement-united-nations-general-assembly

Morrison, S. (2020, September 15). National Energy Address – Tomago, NSW. Prime Minister of Australia's Office. https://www.pm.gov.au/media/national-energy-address-tomago-nsw

Morrison, S. (2021, September 9). Press Conference – Canberra, ACT. Prime Minister of Australia's Office. https://www.pm.gov.au/media/press-conference-canberra-act-23

Mortimer, L., & Park, A. (2014, May 5). Lismore leads nation in adopting solar energy, report finds. *Northern Star*. https://www.northernstar.com.au/news/lismore-leads-nation-adopting-solar-energy-report-/2247827/

Murphy, K. (2020, September 22). Essential poll: two-thirds of voters would prefer Coalition support renewables rather than new gas plants. *The Guardian*. https://www.theguardian.com/australia-news/2020/sep/22/essential-poll-two-thirds-of-voters-would-prefer-coalition-support-renewables-rather-than-new-gas-plants

Narrabri Council. (2020a). 2040 Local Strategic Planning Statement. http://www.narrabri.nsw.gov.au/files/uploaded/file/Planning%20and%20Development/Strategic%20Documents/Local%20Strategic%20Planning%20Statement%202040%20FINAL.pdf

Narrabri Council. (2020b). 2019/2020 Annual Report. https://www.narrabri.nsw.gov.au/files/uploaded/file/Your%20Council/Strategic%20and%20Corporate%20Plans/Annual%20Reports/2019-2020%20Annual%20Reports/Annual%20Report%202019-2020_Final%20w%20Appendices.pdf

Paterson, A. (2017, January 18). 'Power to the people' *The Northern Star*, p. 3. https://www.northernstar.com.au/news/lismores-sustainable-solar-success-story/3132900/

Przepiorka, W., & Liebe, U. (2016). Generosity is a sign of trustworthiness – The punishment of selfishness is not. *Evolution and Human Behavior*, *37*(4), 255–262. https://doi.org/10.1016/j.evolhumbehav.2015.12.003

Rudiak-Gould, P. (2014). Climate change and accusation: Global warming and local blame in a small island state. *Current Anthropology*, *55*(4), 365–386. https://doi.org/10.1086/676969

Santos. (2020a, September 15). Santos welcomes government's focus on unlocking new gas supply. https://narrabrigasproject.com.au/2020/09/santos-welcomes-governments-focus-on-unlocking-new-gas-supply/

Santos. (2020b). Santos welcomes federal signoff on Narrabri Gas Project. November 24. https://narrabrigasproject.com.au/2020/11/santos-welcomes-federal-signoff-on-narrabri-gas-project/

Shorten, B. (2019, January 29). Transcript doorstop Tathra Tuesday. https://www.billshorten.com.au/transcript_doorstop_tathra_tuesday_29_january_2019

Sinn, H.-W. (2012). *The green paradox: A supply-side approach to global warming*. MIT Press.

Smith, C., & Davidson, H. (2014). *The paradox of generosity: Giving we receive, grasping we lose*. Oxford University Press.

Thatcher, M. (1989). Speech to United Nations General Assembly. Margaret Thatcher Foundation. https://www.margaretthatcher.org/document/107817

Tronti, M. (1965). The strategy of refusal. https://la.utexas.edu/users/hcleaver/TrontiRefusal.html

UN Climate. (2021a, February 26). Greater climate ambition urged as initial NDC synthesis report is published. Press Release. United Nations Climate Change. https://unfccc.int/news/greater-climate-ambition-urged-as-initial-ndc-synthesis-report-is-published

UN Environmental Program. (n.d.). Climate justice. LEO Thesaurus. https://leap.unep.org/knowledge/glossary/climate-justice

Vorath, S. (2016, August 15). Lismore community solar farms set for construction as investors pile in. *One step off the Grid*. https://onestepoffthegrid.com.au/lismore-community-solar-farms-set-construction-investors-pile/

Walker, J. (2020). *More heat than life: The origins of our future in the tangled roots of economics, ecology and energy*. Palgrave Macmillan.

Whitham, M. M. (2021). Generalized generosity: How the norm of generalized reciprocity bridges collective forms of social exchange. *American Sociological Review*, *86*(3), 503–531. https://doi.org/10.1177/00031224211007450

Wilkinson, M. (2020). *The carbon club*. Allen & Unwin.

Part III
Mobilisation

Climate camps and environmental movements: impacting the coal industry and practicing 'system change'

Katja Müller

ABSTRACT

'System change not climate change!' has been one of the prominent slogans the climate movement used during its ever growing campaigning against coal mining and the causally linked climate change. This article traces climate camps as a form of protest for a social movement and investigates its concepts. In analysing its intertwinement with particular sites of pollution, the environmental and the left movement, it demonstrates that the movement's broader aims of changing the system and reaching a better life for all might not always be reached, not even in the small scale of its own practices. Yet, as the juxtaposition of concurrent protest action and political, cultural and industrial development shows, climate camps in Germany, at least, have been making an impact, even if the impact is not unanimously called a success or recognized by all stakeholders.

In January 2021, the mining and energy company LEAG published its new coal field plan, setting the seal on the Welzow-Süd II mine extension. Welzow-Süd II, together with Jänschwalde-Nord and the Reichwalde mine extension, was for more than a decade key for continuing lignite mining and coal combustion: only the mine extensions would have guaranteed the continuation of electricity production from coal in the Lausitz region, in the most Eastern parts of Germany, on the border with Poland. Former planning also included the resettlement of more than 800 people for Welzow-Süd II from the village of Proschim. With the announcement of the new plan in 2021, devastating Proschim and resettling its inhabitants was no longer an issue. In its official statements LEAG justified the planning with the requirements of the novel German 'coal combustion termination law' (Kohleverstromungsbeendigungsgesetz), which regulates the orchestrated shut down of all coal fired power plants and coal mines in Germany by 2038.[1] A few years before, in March 2017, LEAG had announced its intention to discontinue expanding the Jänschwalde and Reichwalde mines. This meant that the 900 inhabitants of Kerkwitz, Atterwasch and Grabko would after a decade of being told differently, now have the certainty that they are off the map of resettlement and demolition. In 2017, LEAG argued that the discontinuation decision was made for purely financial reasons, as it would enable the company to securely plan mining for the next 25–30 years and to expand into other areas of the energy and mining sector.[2]

This article investigates the plausibility of the mining company's argument for discontinuing mining on financial and legal grounds alone. It analyses the activism and protests that took place regarding the mine extensions in the Lausitz for more than a decade and starts in its first part with analysing climate camps as a major form. Climate camps have been gaining support as protest strategies in coal dependent regions through continuously bringing climate change on the agenda and combining it with local struggles against relocation. Civil society actors thereby have been influencing energy governance.

This article argues in its second part, that climate movements – also by means of climate camps – have been impacting political decisions as well as socio-cultural notions of coal mining and the environment, and also have an impact on industrial decisions – whether the fossil fuel sector acknowledges this publically or not.

At the same time, climate activists – also drawing on experiences from earlier environmental activism – work not only to succeed in one instance of mitigating climate change but aim at more profound changes of energy, ownership and political systems. With a call for 'system change not climate change', the movement not only demands a reconsideration of political decision making, consumption processes and ownership structures but attempts to put those demands for a better life for all into practice during climate camps. While stretching only over a limited period of time and space, climate camps help to imagine, test and experience these alternative ways of life. However, as this article demonstrates in its third part, the practice of climate camp organization does not always live up to the expectations of system changes but shows the limits of alternative democratic processes.

Climate camps

Climate camps are a comparatively novel phenomenon. While autonomous camps[3] can be traced back for a long time, and camps related to the environmental movement back to the 1960s at least (Feigenbaum et al., 2013), a novel and quickly accelerating climate camp trend developed after the G8 summit in 2005. An initial spark was the climate camp initiated in 2006 next to the Drax coal fired power plant in the UK.[4] It was followed by several camps for climate action in the next few years, taking place often near airports or power stations in the UK as well as in Canada, Denmark, France, Ireland and other countries of predominantly the Global North.

The first climate camp in the Lausitz took place in 2011, 10 years before LEAG announced its new coal field plans. The Lusatian Climate and Energy Camp (short: *Lausitzcamp*) kicked-off as a novel protest form against the continuous expansion of mining, causing the resettlement of people and severe damage to the climate and environment. The *Lausitzcamp* had been set up by people both from within the area – many of them facing resettlement – and by people from larger cities like Berlin or Leipzig. They started a visible and localized protest in an area characterized by lignite mining for more than a hundred years. The Lausitzcamp in 2011 and in the following years lasted for a few days or a week, providing a space with infrastructure for a number of people to sleep, eat, exchange, learn and voice protest against coal mining for energy production and against climate change. The first years saw a limited number (a few dozen to several hundred people) attending and contribution to the camp programme including workshops, round tables, talks, film screenings, excursions and protest actions. Always finding support in local population – providing camp grounds near Proschim, Kerkwitz or other villages, co-organizing or supporting with infrastructure and set-up – the *Lausitzcamp* was also subject to criticism. The vision of a large circus tent surrounded by numerous smaller tents and self-made infrastructure garnered scepticism from passers-by and lack of comprehension of what the 'green freaks' might want. But returning back year after year to different villages in the Lausitz (Müller, 2018), the campers also experienced growing tolerance and many small instances of acceptance, understanding and approval. The year

2014 marked a mile stone by drawing more than 7000 people to the village of Kerkwitz, from where people formed a human chain across the river Neisse to set up a sign against coal mining on both sides of the border. In 2016, the *Lausitzcamp* provided the necessary space and infrastructure for 4000 people near Proschim, allowing the concerted civil disobedience action of entering the Welzow-Süd mine and occupying diggers and railway tracks for more than 48 hours, disturbing the continuous lignite supplies for the adjunct power plant.

Climate camps like the *Lausitzcamp* are a strategy for environmental movements to draw attention to localized, concentrated contributions to climate change. The Lausitz with the three operating coal power plants Jänschwalde, Boxberg and Schwarze Pumpe contributed 54.6 million tons CO_2,[5] equalling 7% of all energy related CO_2 emissions in Germany in 2017, 36% of all lignite-related CO_2 emissions, or 19% of all fossil-energy related CO_2 emissions in Germany in 2017.[6] The Lausitz with its open cut coal mines and power plants hence is a very suitable location for protest: it provides a very tangible, materialized form of climate change appearing as one of its sources. The mines and power plants form a visible stage and precise target for environmental protection and climate change protest. However, being located in a sparsely populated area, protest around open-cut mines or power plants aiming at large participation or a longer term needs a minimum infrastructure of food, shelter and electricity.

Climate camps, as protest camps in general, can provide this infrastructure for large scale protests away from cities and towns. According to Frenzel (2011, p. 164), climate camps are in the first place functional. They are a temporary structure creating a place equipped with the necessities for several day long protests. Yet, this functionality goes hand in hand with advocating more conscious socio-ecological relations, resulting in an ideal of making the camp infrastructure as resource-saving as possible. Put into practice, the Lausitzcamp generated needed electricity from mobile solar panels instead of diesel generators, aimed at saving water, provided sanitary infrastructure in biodegradable form and worked with cooking collectives that prioritized vegan food and rescued groceries. This sustainable take on infrastructure reflects one of the four pillars of contemporary climate camps: to put the vision of a better life into practice. While protesting against larger, systematic exploitation of nature and global warming, climate camps always have the aspiration to demonstrate that a responsible treatment of resources and a severe mitigation of greenhouse gases also comprises sustainable end usage. Beyond socio-ecological relations guaranteeing environmental sustainability, the vision of a better life also encompasses forms of communication and decision making. A better communal life would entail, for instance, nonviolence and consensus-based decision making. The daily plena during climate camps or the ones employed for organizing the camps are two of the spaces where this is put into practice.

A second pillar of protest camps is networking for the movement. While telephone conferences, newsletters or mailing lists are also part of new social movements' networking practices, camps allow for a longer timeframe and face-to-face meetings for networking. Climate camps comprise organizing beforehand – usually in form of plena, in 2020 and 2021 due to the Covid19 pandemic primarily through telephone or video conferences – which allow for the establishment of personal relationships as well as alliances for the cause. The camps themselves provide room for making plans, harmonizing time frames and capacities, forming new alliances or deepening existing ones. Furthermore, connections and acquaintanceships can become more profound through actually living together during the camp and through shared experiences. Networks during climate camps in Germany have been formed between experienced activists from the left and young, ecologically motivated people initially less critical of the wider political and economic system (Kössler, 2013, p. 188). The Lausitzcamp, too, saw a diverse range of experienced activists and newcomers.

They find common ground in being no longer satisfied with the way the political mainstream is treating climate issues, and hence mobilize for climate camps as a place where environmental issues as thematic foci combine with camps as a form of protesting for alternative modes of organizing all aspects of society, comprising last but not least large scale (political) decision making structures. With the rise of activism among even younger people through Fridays for Future, climate camps also faced the questions of how to balance networking and the expansion of climate protest movements and the configuration of the third pillar: action.

Action, as the third pillar, makes climate camps attractive for activists in the more narrow sense of the word. Rather than petitioning, lobbying or campaigning, climate camps have been set up initially to allow for protest action at key locations of environmental pollution, greenhouse gas emissions or climate change impact. Action in the Lausitzcamp, for example, took shape in registered protest marches, human chain, die-ins, occupation of offices, or human-formed pictures. In recent years, groups such as Ende Gelände, ausgeCO2hlt, Sand im Getriebe, Robin Wood or other alliances also conducted civil disobedience action in the context of climate camps. Next to banner dropping and blockades, occupations of open cut mines, coal railways and power plants have gained importance since 2015. These large scale actions are concerted and well-organized. Informing participants about options and (legal) consequences, discussing and formulating consensus about what is appropriate and what is not, forming small and medium sized reference groups, physical training in overcoming potential obstacles, as well as strategically planning routes, timings and logistics have in the past few years led to day long mass occupations of coal infrastructure. Large scale civil disobedience in coal mining areas is a form of action that has not only effectively disrupted coal mining and drawn media attention to the climate crisis coproduced in Germany and Europe, but has required mining companies and police to pay attention and adjust their strategies and handling of protesters. In sum, the mass occupations have resulted in comparatively little civilly and criminal prosecution, given that up to 4000 people entered the private property of coal mining companies.

The fourth and final pillar of protest camps is education. Climate camps are important for educating participants and visitors on environmental and climate related issues through workshops, talks and presentations. Thematically, they can cover everything from upcycling to climbing, from electricity production basics to discussions between activists and mining companies, from kids space to film screenings. Between 2015 and 2019, education at German climate camps also found expression in the degrowth summer school, hosted by the Rhineland and Leipziger Land climate camps.[7]

Climate camps making an impact

The assessments of climate camps' impact on mitigating climate change, and of the Lausitzcamp on abandoning the mine extensions Welzow-Süd II, Jänschwalde-Nord and Reichwalde differ significantly. While the mining company was quick to resort to the claim that economic and legal reasons were alone responsible,[8] the announcement was celebrated as a success by the local activists and the climate movement. In autumn 2017, the local initiatives invited all supporters, especially those from the climate camp, to celebrate and thank them for their engagement that led to this success.

It is nearly impossible to measure the precise impact of environmental movements on a particular political policy or industry's decision, because decision making (in this case for an overall operation plan) is a process, and hence influenced by numerous factors and actors, which are closely intertwined and numerically not graspable. It is clear, however, that the Lausitzcamps drew strong media attention, even making it to the Sunday evening news of national TV.[9] The Lausitzcamps were visited by an ever increasing number of local people, interregional and international activists, and state and national politicians. The activists' continuous efforts were noted by opponents and proponents alike. They drew attention and resources from the mining company, who erected fencing, developed its press and media responses, assigned security guards, mobilized parts of its staff, and filed complaints and brought charges. The mining company had to engage with the movement and its protest forms.

Furthermore, political authorities increasingly paid attention to voiced climate concerns, indicating that the continuous engagement of the activists in the Lausitz has co-produced political change. While it cannot be clearly defined *how much* the climate camps and the climate movement in general contributed to political change, the concurrent strengthening of the German climate movement and the development of fossil fuel critical national politics is striking. The number of people participating in climate camps – as only one indicator of the movement – grew steadily from a few hundred in 2011 to several thousand in the second half of the decade. The rise of Fridays for Future and Extinction Rebellion in 2019 further contributed to climate change becoming a relevant topic in the public sphere and of public engagement, drawing 300,000 people in April 2019 to the streets in Germany alone.[10] At the same time, developments in industry and politics heralded the decline of coal mining: In 2014, the Swedish state company Vattenfall announced its decision to sell the mines and power plants in the Lausitz in the aftermath of the elections of a new parliament. One year later, the German minister for economics announced plans for a *Klimaabgabe* (a climate fee) but substituted this with a *Kapazitätsreserve* (capacity reserve) being a compensation the government pays for shutting down parts of Jänschwalde and other power plants. In 2016, the Czech company EPH bought the mines and power plants, calling the new company LEAG. This year also saw the large Ende Gelände protests at the Lausitzcamp.

Two years later, in 2018, the German government started convening the 'Coal Commission' to work out a plan for ending coal mining and combustion, with Hannelore Wodtke, who had been fighting for the survival of Proschim, as one of its members, as well as three representatives from environmental NGOs. This Commission on Growth, Structural Change and Employment ('Coal Commission') in its report to the government gave its advice on abandoning all coal mining and combustion for power production by 2038, thus effectively shutting down or converting to another energy resource Germany's 104 fully or partially coal-fired plants. Hannelore Wodtke was the only one to vote against the report, as it did not secure the survival of Proschim, but left this decision to LEAG. She did, however, regard fixing an exit date as a success for the environmental movement.[11]

While a connection between the social movement, political legislation and the mining companies' decisions is evident, the LEAG still decided to claim its decisions were based purely on economics. From the movement's point of view, the question of 'success' can likewise not be answered straightforwardly, at least not from everyone's point of view. When the government transformed the Coal Commission's advice in parts into national law and passed the mining areas structural support law and the coal exit law in 2020, this could be termed a success of the movement. Activists achieved their goal of pushing for an exit from coal mining and combustion as one of the largest emitters of CO_2, and succeeded in the political realm when the government passed a law with a precise timetable for shutting down all lignite mining and coal fired power plants. However, not only was Hannelore Wodtke not completely at ease with terming the report and the law a success, but parts of the environmental and climate movement were soon to criticize the legally manifested coal exit as not fast enough and too costly. Energy market mechanisms might have led to an earlier shut down of the power plants and mines, and would not have required the huge subsidies for the shut down or capacity reserves. The set timetable, furthermore, did not correspond to the Coal Commission's advice for timely coal plant deactivation, and the legislation also left room for approving the activation of a new coal-fired power plant. The coal exit law was hence criticized for breaking the consensus that the Coal Commission's diverse stakeholder had accomplished. In sum, there is no univocal approval of the coal exit law as a success of the climate movement, albeit not due to doubts on the movements impact, but because perceptions of what is a 'success' differ, also within the climate movement.

If, as Guigni states (1999, p. xxii), what really matters is the collective benefit to the beneficiary group, and this group is defined as the broader population being spared the emissions from 176 tons of coal from the Jänschwalde-Nord mine,[12] as well as the power plants and other mines discontinuing operation until 2038, then we can speak of a success of the Lausitzcamp and other environmental protest. If we consider the structural and cultural level (Guigni 1999, p. xxiii), we also see success of the movement, as the climate camps in the Lausitz contributed to strengthening solidarity among and beyond the movement, and shifts in public attitudes towards coal mining and combustion being without alternative. The Lausitzcamp, through focussing on and implementing the pillars of action, education and networking, impacted not only political decision making but also had an impact in the industrial and socio-cultural realm, within which coal mining in the Lausitz was for a long time most highly regarded, both economically as well as regards identity politics (see Goodman et al., 2020, p. 116ff.). However, if we define the beneficiary group as the villagers of Proschim not having to leave their homes, the climate movement has not succeeded.

Neither is the world's most vulnerable population or humankind saved from climate change. Even if the passing of the coal exit law – or the even 'smaller' preceding success of stopping the Jänschwalde-North mine extension in 2017 – is regarded as a success, it can be read as an example of the climate movement winning the battle but losing the war. It is an instance of tackling climate change and contributing to its mitigation, but not stopping it altogether or even limiting global warming to 1.5°C. The climate movement thus echoes the development of the environmental movement, of which it can be regarded as a part or as emerging from.[13] The environmental movement was, by the end of the 1990s regarded as a successful social movement (Rucht, 1999): it had grown and become consolidated, it had influenced politics, society and culture. The movement had been diversified and by the turn of the millennium had covered almost every aspect of environmental protection (Rucht, 1999, p. 205). This growth and diversification was not only in breadth but also in depth: environmental activism had shown a growing professionalism and expertise. Greenpeace is regarded as a prime example for this professionalization and institutionalization (Zelko, 2014), although institutionalization and structuring activism also comes at a cost (Mildner, 2002). For the environmental movement's impact on society and politics, it can be stated that

> environmentalism is highly valued. Hardly any relevant social group, hardly any important political party, can afford to reject the goal of environmental protection. Even those industrial branches and corporations which, on concrete issues, act as opponents of environmental groups rhetorically embrace environmentalism and try to gain a 'green and clean' image. (Rucht, 1999, p. 206)

In Germany, this had seen not only the introduction of environmental issues into policy making and legislation, or the Green party's participation in governing the country from 1998 to 2005, but the introduction of multiple environmental agencies, state departments and advisory boards, educational programmes, climate protection plans and rules and regulations. At the same time, environmental movements had been impacting the behaviour and attitudes of people. Making environmental issues heard and becoming a relevant voice had also resulted in a striking level of attitudinal support for environmental protection (Rucht, 1999). This comparatively swift gain in popularity and take up by the population is another clear indicator of the movement's success.

This situation is comparable to the climate movement, which – as described above – also found its way into governmental planning and politics, public opinion, and positive attitudes, with the climate camps being but one instance of it. However, the dilemma remains as regards environment and climate: the situation of the environment in total has not improved over the last fifty to sixty years, neither has that of the climate. Quite on the contrary, it has become worse. Extractivism and the exploitation of nature went along with environmental destruction that resulted in a degree of pollution that no longer affects only immediate surroundings but climatic conditions, which again negatively affect the environment. The climate movement, and the environmental movement in more general, has hence not been succeeding in its larger aim of mitigating climate change or effectively protecting the environment overall; it has been influencing single sites, but not making enough impact. Instances of halting or changing pollution at particular sites cannot outweigh the overall destruction still being done.

Climate camps can be seen as acknowledging this problem and working towards its solution. Rather than tackling only one issue, they demand 'system change not climate change', and try to put this into practice in small scale during the camps. Climate camps also developed in close relation to the political left scene (Kössler, 2013, p. 188), with activist from attac, Antifa and other left groups undermining the notion of an effective protection of environment and climate needing more than fighting single power plants or airports – albeit acknowledging such sites as very suitable instances for manifesting protest. Before and after the G8 protests in Heiligendamm in 2007, left groups' coordination meetings put climate as an issue on their protest agenda, with camps being perceived as a suitable form for this work. The precise disentanglement of originators of the climate camps – starting in the UK in 2006 and in Germany in 2010 – within a movement as loosely organized as the climate movement and overlapping in parts with the left scene, is hardly possible. In practice, climate camps draw on experiences, values and agendas generated previously in both the left and environmental movements. As Woodsworth (2008, p. 22), focussing on the UK, states,

> From the 1960s new Left and student movements come the roots of counterculture and alternative notions of community [to climate camps]. From second-wave feminism, at its height in the 1970s, comes cultures of political organising that seek to avoid hierarchy and oppression. The 1980s brought the peace/anti-nuclear movement and its use of non-violent direct action and protest camps as well as encounters with new countercultural currents in the form of New-Age travellers and anarcho-punks, and the more militant tactics of animal rights activists. The 1990s saw the consolidation of a specifically environmental direct action movement resisting road-building and later genetic modification, and the rise of anti-capitalist/alter-globalisation/global justice (ACAG) movements [...]

The intertwinement with leftist groups and justice movements brought the long standing tradition and profound analyses of current and desirable socio-ecological relations to climate camps, which see reasons for the current climate crisis in capitalism, technology or culture (Bebbington, 2017, p. 3), which need to be changed.

To bring about these changes needed for effectively protecting the climate is a mammoth task. It is not clear if the movement will succeed or fail, and what means are the best to make the necessary impact. During organizing or discussing climate camps and other forms of activism, moments arise where the effectiveness, impact and sense of environmental action is questioned. Activists discuss and scrutinize the means applied – whether disruptive or moderate forms are more successful, if applying disruptive tactics like Ende Gelände's blockades of mining infrastructure on the occasion of climate camps are more productive than participating in the Coal Commission. It has been argued academically that 'disruption is the most powerful resource that movements have at their disposal to reach their goals, since they lack the institutional resources possessed by other actors, such as political parties and interest groups' (Guigni, 1999, p. xvii), and that a movement's political context, the available political opportunities and the repressiveness of the rulers, as well as the cultural climate and the movement's ability to create innovative and disruptive tactics are essential for disruption to be more or less effective (Guigni, 1999). To put (disruptive) activism into practice and exercise it within a given socio-cultural and political context is, however, over and over again subject to trial.

Changing the system – not even within climate camp organization?

This also goes for small scale systematic changes, which are put into practice during climate camps as a way to establish a communication and decision making processes that are free from violence and hierarchies. Of central importance is consensus, meaning that decisions are not based on majority votes, but that agreements are reached among a broad proportion of people involved (Bressen, 2007, p. 213). Practicing basic democratic and consensus oriented processes is an essential part of the movement's conception (Burkhart et al., 2017, p. 110), as it is a more democratic decision making for real political participation. Proposals can be discussed and amended several times, concerns heard and options pondered, while the discussion rules aspire equal shares of contribution by diverse individuals (for example through rearranging the speakers' list, favouring those who until then spoke less). Voting eventually comprises articulation of compliance, small or large concern, 'stepping aside' or a veto – with the latter being the only form of blocking a proposal, and the former being various forms of acceptance. These modes and rules guarantee that the largest possible numbers rather than a simple majority supports a decision, contributing to the levelling of hierarchies and the application of grass-roots democracy. Working as decentralized, periodically reorganizing networks of people becoming active for climate change mitigation, climate activists put these modes of decision making into practice when bringing together individuals, local initiatives, branches or groups. All of them have equal rights and become part of what is envisioned as horizontal decision making.

Two instances from climate camp organizing, however illustrate the limits in practice of this grass-roots democracy. In 2018, after holding the Lausitzcamp for seven consecutive years and the announcement of the mining company to discontinue the mine-extension plans for Jänschwalde-Nord and Welzow Süd II, organizers of the Lausitzcamp in consultation with other climate movement groups decided for a change of locations and a shift towards the so called Central German mining area (Mitteldeutsches Revier – after the Rhineland and Lusatia the third lignite mining area in Germany). In February 2018, the first organizational plenum took place in a former factory complex in Leipzig.

Half a dozen of the Lausitzcamp organizers were present, as were organizers of the degrowth summer school and about 30–40 people who had not previously been involved in organizing but followed the call to organize a camp in 2018 somewhere around Leipzig and the Central German mining area. After warm-ups and first introductions, someone from the Lausitzcamp introduced the camp's history, its development, guidelines and present with a slideshow. Everyone got a visual impression of the Lausitzcamp and a rough idea about its trajectory and the climate camp's central ideas. After the person organizing the Lausitzcamp for the last year had finished, a person previously organizing the summer school rose to speak, saying that the Lausitzcamp had been cancelled for 2018 and that the group is instead liaising with the summer school. The

Lausitzcamp people immediately objected, stating that the Lausitzcamp had by no means been cancelled, but was going afield, carrying the anti-coal protest to another mining area but keeping its distinct name and drawing on gathered experience and support. They also declared that while the Lausitzcamp was to go afield in 2018, its return to the Lausitz in 2019 is important, because the mining company scheduled its decision about the village Proschim for 2020.

This different understanding of who supports whom, of who becomes an ally or part of another group turned out to be an issue of importance over the next two days. Such questions of structure – where does one movement end and another one start, who is a partner, who is an ally? – have long been a characteristic of climate change movements (Bebbington, 2017). In this particular case, the question was answered through a combination of planning and moderating the two and a half plenary days with discussions and decision making of the whole group – less so through discussing the issue of cooperation and eventually reaching an agreement among the largest possible number of people. People eventually opted for a new climate camp with a completely new name – Klimacamp Leipziger Land – in consequence of which a few of the Lausitzcamp organizers withdrew and dropped out from organizing the climate camp. Others, however, stayed on, setting up a week-long climate camp located next to the open cut mine Vereinigtes Schleenhain in the village of Pödelwitz, drawing more than 1000 campers to the workshops, protests actions, performances and summer school.

Two years later, at the beginning of 2020, that year's kick-off organizational plenum for the Klimacamp Leipziger Land showed again how organizing a meeting and moderating it can undermine the ideal of discussion based agreement and consensus. This plenum took place in an open office of a member of parliament, and about two dozen people gathered. The inhabitants of Pödelwitz had announced that after hosting the climate camp for two consecutive years with a very positive reception and feedback within the village and beyond, they would prefer the 2020 edition of the climate camp to take place elsewhere.[14] Hence, the plenum's main task was to decide where, when and with what thematic subheading the 2020 climate camp would take place. The plenum started with the by now familiar warm-ups and rounds of introduction, followed by the task to discuss in small groups what grass-roots democracy comprises – a result of criticism in its fragmentary implementation in previous years. Afterwards, ideas for locations and topics were collected, groups formed to think through these ideas and posters drawn for each one of them, which the respective small groups later explained to the whole group. The ideas ranged from a camp to rejuvenate a rundown riding centre in the woods slightly outside of Leipzig, to tying up with the airport protest in a nearby small town, to gathering all over the city on the occasion of the transport summit scheduled for May 2020, to carrying the protest to a small town in the hinterlands known for right-wing and climate change denial tendencies. Pros and cons were exchanged on the different scenarios, ranging from logistics to early dates, from missing camp character to expected engagement from visitors vs. their 'service attitudes'. On the second day, the moderator dismissed two of the seven suggestions from the start, saying that she spoke with the larger climate change network and that dates would collide with other protest action. In the course of the day, it showed again and again that moderation is key to the decision making process – and that despite the initial discussion of what grass-root democracy comprises, it is not put into effective practice. This became most obvious when out of the remaining three ideas, only two were discussed in detail while one was put aside by the moderator without reason beyond making room for an opinion poll on the other two. The moderator met the slight objections to this procedure by pointing out that this was only to get an idea of the opinion but not a decision making. When everyone was asked to get up and position themselves physically on an imagined line of support of idea one – the traffic summit in May – wording made clear that when you stand at the very end, you'd be out of the process more or less, signalling 'that you can't support this idea'. While there was one abstention, the majority – being bodily also subject to group dynamics – positioned themselves towards being supportive of the idea. Slightly less people supported the second idea, and the third one still in the running was 'now that there is a clear poll' neither discussed not balloted.

These two instances illustrate a slight discrepancy between ideal and practice of alternative democratic processes. Climate camps feed upon the

> uptake of the complex normative ideas that underpin contemporary environmentalism – about how decisions should be made and by whom, about holism and systems thinking, about the nature of progress and its relationship to economic growth, and about the relationships between nature and culture, local and global, present and future generations. (Woodsworth, 2008, p. 18)

A socio-ecological transition for climate change mitigation and social and ecological justice must comprise overcoming the growth paradigm and destructive economics. Better economic systems and societies will rest on values like attentiveness, solidarity and cooperation (Burkhart et al., 2017, p. 14), which are to be demanded as general principles on multiple action scales as well as to be put into practice in one's own acting. For climate camps as part of the climate movements, this implies that decisions should be made in a nonviolent and consensus-based form. Small-group discussions, delegate systems, plenary decisions with the abovementioned voting options and a speaking order attentive to shares in discussions and marginalized speakers are to provide for non-hierarchical and grass-roots decision making.

However, the two organizational plena illustrate that the implementation of these values poses challenges. While people comply with the rules of voting, and display attentiveness and solidarity, the basis democratic ideals can be undermined by moderation and agenda setting. The organizers and moderators of a plenum are necessary as facilitators, but they also have the advantage of setting up the agenda and driving or directing things. Setting up plena requires thinking things through in advance, making suggestions about forms and topics that will make sense to successfully set up a climate camp (or reach any other envisioned outcome). It requires a careful framing that allows the process to advance, but at the same time needs to be very open to the ideas and needs of the whole group. The group usually entrusts the 'processors' and moderators with facilitating the process. In their role as moderators or 'processors' they are to intervene from an intersubjective point of view only, without being biased. Yet, given that objectivity is not possible and that in many cases the moderators and processors are not external, their action can – and should in some instances, as for example when they give so far marginalized people more room – privilege some participants and disadvantage others. What is legitimate in a process aiming at levelling pre-existing inequalities, becomes problematic when the moderator's interference becomes an acting for her own interests and turns into bending the established rules and overstepping previously set lines. The unconscious or skilled setting of the agenda, of assigning topics and tasks, of scheduling opinion polls or votes or delaying the same allows both moderators and processor to steer the process. It becomes impossible to draw clear-cut lines between facilitating, operating, regulating and controlling the discussions and decision making process.

It also needs to be acknowledged that in the particular cases of organizing the climate camp Leipziger Land some participants were professionally engaged in degrowth or environmental activism, while others were contributing to setting up the camp in their free-time. It goes without saying that the former had a clear advantage in resource allocation, resulting in investing more time in planning and controlling and having more information at their disposal. Having the resources to be a delegate and/or keep oneself constantly informed about other groups' or the larger movement's strategies and plans, enlarges not only the capacity to use and integrate this information and to contribute to a more fruitful and informed planning, but also increases symbolic capital and can raise the status within the group. In some cases, information can be used to achieve one's own aims and in the process might undermine the larger values of solidarity, attentiveness and cooperation that the movement wants to rest upon.

Conclusion

'System change not climate change!' has been one of the prominent slogans the German climate movement has used during its ongoing campaigning against coal mining and the causally linked climate change. The idea that the current extractive capitalism is the wrong political context for an economy that takes environment into account and that a general rethinking of politico-economic conditions feeding on growth and maximizing profit derives in parts from the intertwinement with left activism, and in parts from experiences of 'failure' in protecting the environment at large. This notion is interrelated with grass-roots democratic practices and protest forms targeting tangible manifestations of extractivism, for which climate camps form both a functional setting and a small scale space where ideals of a 'better life for all' are put into practice. This article has shown that organizing and moderating decision making processes for climate camps does not always live up to the grass-roots democratic standards set for and by themselves and that internal decision making processes are not as non-hierarchical as imagined or proclaimed, but continue to comprise imbalances. However, social movements are often faced with the questions of 'success' or 'failure' of set goal, targets and ambitions, without being able to give an ultimate answer. Because what someone within a movement as loosely organized as the climate one terms a failure, can be read by another person as a step forward or success. What this article has shown, however, is that the climate movement with camps as one of its most prominent forms has had an impact on coal mining in Germany, not only as regards the mine extensions in the Lausitz, but also as regards the socio-cultural framing, policy making and industrial decision making – which is but one step towards keeping global warming below 1.5°C.

Notes

1. https://www.zeit.de/news/2021-01/14/leag-legt-plan-vor-welzow-ii-ist-raus-und-proschim-bleibt.
2. http://www.niederlausitz-aktuell.de/niederlausitz/66170/leag-gibt-plaene-fuer-lausitzer-tagebaue-weitgehend-auf.html.
3. See Feigenbaum et al. (2013) on the division of camps.
4. https://web.archive.org/web/20070301045057/http://news.independent.co.uk/environment/article1223134.ece.
5. https://www.thru.de/daten/.
6. Totalling 735.1 million tons in 2017: https://www.umweltbundesamt.de/daten/klima/treibhausgas-emissionen-in-deutschland#treibhausgas-emissionen-nach-kategorien; 149 million tons CO_2 emissions due to lignite, 283 million tons due to fossil energy production: https://www.umweltbundesamt.de/sites/default/files/medien/384/bilder/dateien/6_datentabelle-zur-abb_entw-co2-emi_2020-03-11.pdf.
7. In 2010, the first official German climate camp took place in the Rhineland mining area in Western Germany, about 40 km from Cologne and Dusseldorf. It focussed on the large scale open cut mines and the adjunct power plants as a tangible and visible target for climate action. The Lausitzcamp followed one year later. The Lausitzcamp has been taking a break since 2017, but saw a single day actions in late 2019. The latest climate camp to be formed was the Klimacamp Leipziger Land (Climate Camp Leipzig hinterland), taking place in 2018 for the first time. In 2020, due to the Covid19 pandemic, the Klimacamp Leipziger Land was conducted online, and the Rhineland camp did not take place. Instead, a large scale Ende Gelände protest weekend took place in the Rhineland in September, which drew on

existing personnel and material infrastructure, but relied on decentralized, smaller spaces for spending the nights as but one aspect of a Covid19-related safety concept.

8. https://www.lr-online.de/leag-aendert-tagebauplaene-39724043.html.
9. https://www.tagesschau.de/multimedia/sendung/ts-14035.html.
10. https://fridaysforfuture.org/what-we-do/strike-statistics/list-of-countries/.
11. https://www.maz-online.de/Brandenburg/Hannelore-Wodtke-stimmte-als-einzige-gegen-den-Abschlussbericht-der-Kohlekommission.
12. According to Tudeshki et al. (2007, p. 14).
13. Environmental and climate movements are sometimes seen as one and the same thing or the latter an advancement of the further. It took a while for climate change and global ecology to be a prominent issue of protest, while modern environmentalism in the Global North took off in the 1970s already, often along lines of nature conservation and political ecology (see Giugni & Grasso, 2015). Precise cases of species or places of nature being threatened were on the agenda of environmental movements as well as broader issues of biodiversity loss, deforestation or acid rain. In the Federal Republic of Germany (FRG), the environmental movement turned in the mid-1970s to nuclear power, making anti-nuclear protests the major line of protest for the coming decades. While environmentalism in the FRG is said to have been quite radical in its beginnings and becoming to some extent institutionalized or turned mainstream during the 1980s (Rucht & Roose, 2003), other countries in the Global North experienced an up and down in the 1980s (Woodsworth, 2008, p. 18). There has been constant academic debate about whether the environmental movement has been declining or rising since the 1970s (see Russell, 2015; Woodsworth, 2008, p. 18), especially since it is difficult to grasp a rather loosely organized movement, defined only by having an agenda with some relationship to a politicized conception of nature and some sustained forms of collective action (Bebbington, 2017, p. 1). However, what can be noted is that global ecology and climate change became stronger issues for the environmental movements only in the 1990s, arguably after the Rio Earth Summit in 1992 (Giugni & Grasso, 2015, p. 340f.) and in relation to CFCs in the atmosphere (Roose 2012). The strategies to draw attention to environmental issues, to increase pressure on political decision makers and polluters, and to change the conditions for the better, reach from situational protest and demonstrations to petitions, lobbying and campaigning. Protest camps, too, have been part and parcel of some protests, yet emerged anew for climate change mitigation and climate direct action in the late 2000s.
14. The two climate camps in Pödelwitz raised attention to the village's fate and contributed to all three partners of the newly elected government of the state of Saxony to agree on the preservation of the village. In late 2019, this preservation found its way into the coalition contract. The so called Coal Exit Law passed in 2020 underlined this agreement, and in January 2021 the mining company, after discussions with the ministry of economics, drafted a renewed mining plan, providing the needed legal ground for Pödelwitz to be off the devastation map.

Disclosure statement

No potential conflict of interest was reported by the author(s).

ORCID

Katja Müller http://orcid.org/0000-0002-9465-1802

References

Bebbington, A. (2017). Environmental movements and protest. In D. Richardson, N. Castree, M. F. Goodchild, A. Kobayashi, W. Liu, & R. A. Marston (Eds.), *The international encyclopedia of geography*. Wiley. https://doi.org/10.1002/9781118786352.wbieg1090

Bressen, T. (2007). Consensus decision making. In P. Holman, T. Devane, S. Cady, & Associates (Eds.), *The change handbook: The definitive resource on today's best methods for engaging whole systems* (pp. 212–217). Berrett-Koehler.

Burkhart, C., Schmelzer, M. & Treu, N. (Eds.) (2017). *Degrowth in Bewegung(en)*. Oekom Verlag.

Feigenbaum, A., Frenzel, F., & McCurdy, P. (2013). *Protest camps*. Zed Books.

Frenzel, F. (2011). Entlegende Ort in der Mitte der Gesellschaft Die Geschichte der britischen Klimacamps. In A. Brunnengräber (Ed.), *Zivilisierung des Klimaregimes: NGOs und soziale Bewegungen in der nationalen, europäischen und internationalen Klimapolitik* (pp. 163–186). VS Verlag für Sozialwissenschaften.

Giugni, M., & Grasso, M. (2015). Environmental movements in advanced industrial democracies: Heterogeneity, transformation, and institutionalization. *Annual Review of Environment and Resources*, *40*(1), 337–361. https://doi.org/10.1146/annurev-environ-102014-021327

Goodman, J., Connor, L., Ghosh, D., Kohli, K., Marshall, J. P., Menon, M., Müller, K., Morton, T., Pearse, R., & Rosewarne, S. (2020). *Beyond the coal rush. A turning point for global energy and climate policy?* Cambridge University Press.

Guigni, M. (1999). Introduction. In M. Guigni, D. McAdam, & C. Tilly (Eds.), *How social movements matter* (pp. xiii–xxxiii). University of Minnesota Press.

Kössler, G. (2013). Die Klimabewegung in Deutschland. In M. Dietz (Ed.), *Die internationale Klimabewegung* (pp. 185–204). Springer.

Mildner, G. (2002). Action follows structure vs. structure follows action? (Bewegungs-)Organisationen und Strukturation. In S. Bartmann, K. Gille, & S. Haunss (Eds.), *Kollektives Handeln* (pp. 163–187). Hans-Böckler-Stiftung.

Müller, K. (2018). Mining, time and protest: Dealing with waiting in German coal mine planning. *The Extractive Industries and Society*, *6*(1), 1–7. https://doi.org/10.1016/j.exis.2018.09.001

Roose, J. (2012). Wollen die Deutschen das Klima retten? Mobilisierung, Einstellungen und Handlungen zum Klimaschutz. *Forschungsjournal soziale Bewegungen*, *25*(2), 89–100. https://doi.org/10.1515/fjsb-2012-0215

Rucht, D. (1999). The impact of environmental movements in Western societies. In M. Guigni, D. McAdam, & C. Tilly (Eds.), *How social movements matter* (pp. 204–224). University of Minnesota Press.

Rucht, D., & Roose, J. (2003). Germany. In C. Rootes (Ed.), *Environmental protest in Western Europe* (pp. 80–108). Oxford University Press.

Russell, J. D. (2015). Waxing or waning? The changing patterns of environmental activism. *Environmental Politics*, *24*(4), 530–552. https://doi.org/10.1080/09644016.2015.1023576

Tudeshki, H., Könnecke, M., & Rebehn, T. (2007). *Studie zur Fortschreibung der Tagebauentwicklung im Lausitzer Braunkohlerevier (Teil Brandenburg)*. Retrieved May 5, 2021, from https://lbgr.brandenburg.de/media_fast/4055/Braunkohlenstudie_Mai2007.pdf

Woodsworth, A. (2008). *Growth in the UK climate direct action movement: Experience, politics and practice* [PhD thesis]. University of East Anglia. Retrieved May 5, 2021, from https://core.ac.uk/download/pdf/40006868.pdf

Zelko, F. (2014). *Greenpeace: Von der Hippiebewegung zum Ökokonzern*. Vandenhoeck & Ruprecht.

Climate movements in Germany, India, and Australia: dynamics of transition, transformation, and emergency

James Goodman and Tom Morton

ABSTRACT

Climate movements have emerged with increasing force as governments continually fail to address the mounting climate crisis. Successive movement strategies have proved integral to the possibility of effective climate policy and to wider prospects for intentional climate agency. Tracing the emergence of climate movements from the mid-2000s, this article highlights three main developments in strategy. First climate 'transition' centres on policy advocacy and socio-technical solutions. Second, efforts at climate 'transformation' signal a move to grassroots organizing for just alternatives against fossil fuel dependence. Third, climate 'emergency' centres on the mobilization of symbolic and disruptive outrage at present and future climate change impacts. Dynamics on the ground are explored through analysis of developments in Germany, India, and Australia, as contrasting post-industrial, post-colonial, and extractivist cases. Analysis of climate movements across the three contexts draws-out converging themes and possibilities for broad-based cross-sector mobilization.

Climate movements have become highly generative sites of social transformation, yet ironically are born from failure. The annual cycle of hope and disappointment in global climate policy was engrained into climate mobilization from the 1990s. The 1992 UN Conference on Environment and Development that agreed the Framework Convention on Climate Change was itself a focus for large protests, and activists have targeted the annual Conference of the Parties ever since (Hudson, 2015). The twenty-fifth annual COP, held in Glasgow in 2021, conformed to the familiar pattern. Speaking at a protest outside the summit, Greta Thunberg described it as a 'global north greenwash festival' and a 'two-week-long celebration of blah blah blah' (BBC, 2021, november 5. That particular summit failed to meet two of its key goals; a renewal of emissions reduction targets for 2030 that would limit global heating to 1.5°, and a commitment to accelerate the phase-out of coal-fired power (Hales & Mackey, 2021). Representatives of Global South countries described the conference as a fraud; Sanjay Vashist, director of Climate Action Network South Asia, said the Global South had been cheated once again, and the outcome of the conference would 'ensure genocide by extreme weather events in developing countries' (Illius, 2021). In his closing address, the summit president Alok Sharma apologized for the last-minute changes to the agreement text – from a 'phase-out' to a 'phase-down' of coal – and said he understood the deep disappointment of many delegates (Reuters, 2021).

The weakness of the global climate regime gave climate activism its initial purpose. But the role of acting as 'loyal opposition' to global climate policy-makers was inherently self-limiting. As Bullard and Muller argued in 2012, 'global regulation generated global contestation', but this meant a constrained movement, where climate activism could 'only ever be as strong as our antagonist' (Bullard & Müller, 2012, p. 56). Climate activism had moved lockstep with global climate policy, focusing its claims on the regime as a dialectical hinge. Following the 2009 Copenhagen COP, which failed to agree a successor to the Kyoto Protocol, there was a deliberate move to break that nexus. Disillusioned with pursuing abstracted global policy climate activists sought-out more concrete and material sites of leverage (Bullard & Müller, 2012). A more grounded and autonomous climate movement began to emerge, one more capable of defining its own agendas, with its own political targets and horizons. Here, climate activism morphed into the climate movement. Like all social movements, the climate movement has needed to generate a sustained process of collective consciousness and action. It has needed to develop its own organizational form, and its own agenda, to challenge existing institutions and seek to reconstitute the social order. In this, as Tarrow argues, social movements gain the capacity to recreate society, as a 'movement society' (Tarrow, 2011). Climate movements likewise can present a wide-scale challenge to the social order, posing the prospect specifically of *climate* 'movement society' (Goodman, 2022). The question is how they seek to do this – the question of strategy.

Many observers date the growing climate movement from the mid 2000s, suggesting a sustained movement had begun to emerge at this time centred on climate justice. As outlined below, a wave of movement-orientated research has charted cycles of mobilization, how they have cascaded across social fields, and the extent to which the movement has itself become a generative social force. With growing climate impacts, there was a widely-cited upsurge in the scale of climate mobilizations pre-COVID, estimated a rise from 1 million participants in 2015 to over 6 million in 2019 (de Moor et al., 2020). There was also expanded scope, for seemingly 'impossible' solutions (Davis, 2010, p. 45). An ecological social equality agenda emerged, as a 'struggle for ecological sustainability and substantive equality aimed at resurrecting the global commons' (Bellamy-Foster, 2019, p. 14). Equally significant there was a spatial extension, across vastly different contexts, including for instance in India (Ylä-Anttila & Swarnakar, 2017). There are now active climate movements outside the established industrial 'heartlands', in many cases intensified by on-the-ground climate crisis (Tokar & Gilbertson, 2020).

Across these aspects, there is sharp internal debate and contestation between movement organizations over preferred models for climate action. With deepened social immersion in the crisis there has been a dynamic and urgent process of contention over strategy. This article focuses on this process, exploring the different strategies adopted by climate movements. It advances a three-part model for interpreting the emerging movement agendas, distinguishing three modes of mobilization – climate transition, climate transformation, and climate emergency. It develops this three-part model by comparing the agendas of climate movement organizations in contrasting liberal democratic states – India, Germany, and Australia.

Models for climate movement: transition, transformation, and emergency

Climate movements are interpreted in a range of ways by social movement researchers. There is often a distinction drawn between reformers and challengers. Dietz and Garrelts for instance identified a two-part divide between climate modernizers and climate justice advocates (2014, p. 2). Jamison developed a three-part distinction between 'old' forms often expressed in party-political linkages, 'new' environmental movements institutionalized into ENGOs, and the emergent climate justice orientations (2010). Reitan and Gibson also suggested a three-part framework, centred on ideological divides between reformist ENGOs, Marxists, and anarchists (2012). More recently Tokar and Gilbertson found a confluence of three forces: mobilizations against fossil fuel extractivism, environmental justice organizing, and global justice framings: they highlight the situated origins of these aspects, across North and South, and the challenges of advancing climate justice claim under neoliberalism (2020).

The approach taken here focuses on three contending pathways (these mirror distinctions made in Harvey 1996, p. 149). Climate advocacy is dominated by ecological modernization, with hopes resting on a *Transition* to decarbonize economic growth (Mol & Sonnenfeld, 2000). In contrast, *Transformation*-centred approaches take a 'system critical' stance on climate change and develop a range of anti-systemic approaches to address it. More conservation-based approaches assert natural limits and focus on planetary exhaustion as an *Emergency*. The three approaches define very different strategies and agendas for climate movements.

(i) A Movement for Transition

The dominant way of interpreting climate advocacy is in terms of its capacity to secure policy change: this focuses attention on the political process and assumes a broadly pluralist frame and an instrumental approach. Here, the role of the movement is to create and exploit political opportunities, such as through a process of 'frame shifting' for renewables, eco-pricing, and 'green' states (Moll & Sonnenfeld, 2000). There are extensive debates between 'weak' and 'strong' transition, between market-based models for 'carbon pricing' or for divestment on the basis of 'transition risk' assessments, and the more interventionist approaches expressed in the **some of the 'New Deal' proposals for instance** (Aranof et al., 2019). In this context climate science is used as a political resource, a form of populism for climate transition (North, 2011), bolstered by arguments that decarbonization offers economic opportunities. Analysis from this perspective can reveal the generative role of climate advocacy, in both driving the transition agendas and exploiting them for political leverage. A study of climate transition frame-shifting emphasized the new mobilizing resources that can be created (Della Porta & Parks, 2014). From this, climate issues can become a stake in party political rivalry, allowing movement organizations to exploit the 'climate change policy window' (Nulman, 2015, p. 89). Networks of contention, created initially by environmental Non-Government Organizations (NGOs), are seen as central to advancing the transition agenda (Hadden, 2015). There is a stress on mobilizing existing member-based advocacy organizations into 'peak' networks rather than building new movement organizations: with this, climate action emerges from a shared 'interpretative schema' across the existing field of NGOs (Brick & Cawley, 2008). One study by the Global Climate Action Network, for instance, stressed this process of cross-sector and cross-national NGO alliance-building in advancing climate policy (Lipshutz & McKendry, 2011).

(ii) A Movement for Transformation

A second approach is to seek wider socio-ecological transformations as a precondition for addressing the climate crisis, as expressed in the climate justice critique (Chatterton et al., 2012). The approach has an 'anti-systemic' logic dating from the 2010 'Conference on Climate Change and Mother Earth's Rights', the Southern alternative to the failed UN Copenhagen Summit (see

Wallerstein, 2011). A manifesto of sorts for the approach emerged in 2012 with the synthesis document 'Another Future is Possible' at the Rio + 20 Earth Summit (Goodman & Salleh, 2013), with similar orientations developed at the 2014 UN climate summit in New York (Giacomini & Turner, 2015) and again at the 2015 Paris Summit (Cabello & Gilbertson, 2015). Subsequently, the demands have become more domesticated, for instance with versions of the 'Green New Deal' (Aranof et al., 2019). These can be paired with post-growth models, centred on eco-sufficiency (De Young & Princen, 2012) and forms of 'de-growth' (D'Alisa et al., 2010), or with eco-socialist accounts in which capitalist expansion creates its own ecological limits, forcing wider social transformation (Bellamy-Foster, 2002; Kovel, 2007). Several point to 'the commons' as the foundation for ecocentric values (Giacomini & Turner, 2015), and to 'strategic commoning' as a key means of advancing decarbonization (Dahlin & Fredriksson, 2017). There is often an emphasis on planning, for instance as a 'collectively managed, just process' (Mueller & Passakis, 2009, p. 59), including analysis of trade union agendas for 'energy democracy' (Hampton, 2015). Others, from ecofeminism, advance a 'subsistence perspective' centred on living ecologies and forms of ecologically regenerative labour, beyond 'wage labour' (Bennholdt-Thomsen & Mies, 1999, p. 177). The focus on the conditions for social reproduction then offers the potential for inter-sectional alliance-building, a form of 'living environmentalism' (Di Chiro, 2008).

(iii) A Movement for Emergency

More recently, with growing climate change impacts, a wave of climate movement organizations have focused on 'emergency' action, defining climate change as an existential threat and addressing immediate impacts (Foust & Murphy, 2009). From this perspective, the crisis signals humanity's confrontation with its ecological context, akin to watching a 'disaster movie' (Luke, 2014). The approach is highly normative and asserts the need to live within limits, as an imperative above politics, described as a 'post-political' appeal to humanity (Spaiser et al., 2022; Swyngedouw, 2010). The 'emergency' perspective centres on the state but the demands are often not elaborated beyond the need to take action in conformity with climate science (de Moor et al., 2020; Marquardt, 2020). The need to act against ecological exhaustion translates into a strong identification and personalized commitment (Martiskainen et al., 2020). Here, climate movements can take a more expressive form, with prefigurative action linked to forms of symbolic action, from online campaigns to street protests, vigils, and blockades. Through campaigns centred on expressing outrage, including through civil disobedience, emergency movements have created newly mobilized political communities (Saunders et al., 2020). The 'Extinction Rebellion' and 'Fridays for the Future' protests from 2018 to 2020, are good examples of this approach. There is a direct appeal to government, to 'tell the truth' and declare 'climate emergency'; there is no programme beyond this, other than to establish broad-based 'citizens assemblies', as deliberative structures to advance the agenda (Harding, 2020). But there is also an avoidance of an explicit programme, for instance, to address capitalism and the 'political economy of climate change' (Berglund & Schmidt, 2020, p. 100). There are diverse outcomes: one study in Sweden found that apocalyptic framing created political controversy, producing new polarities and antagonisms in the political process (Anshelm & Hultman, 2015). Others have stressed the generative aspect: Beck for instance argued 'anthropological shock' can generate what he called 'emancipatory catastrophism', where despair and outrage become vehicles for action (2015, p. 79).

Comparative analysis

Climate movements develop unevenly, but there may be signs of convergence over strategic issues. Here we seek common themes across India, Germany, and Australia. The analysis applies the three-part model, investigating alignments with 'transition', 'transformative', and 'emergency' agendas, stressing debates between them. All three countries are liberal democracies which officially are seeking a transition process to decarbonize, but continue to expand fossil fuel extraction and use. There are profound differences: Germany is a post-industrial society, India is an industrializing post-colonial society, and Australia is an extractivist settler society. Germany has a strong national commitment to climate action, with intense debates over how to achieve ambitious targets. National ambivalence prevails in Australia, with climate policy heavily politicized for over a decade, in contrast with extensive State government renewables initiatives. In India the national government has pursued emissions reduction, mainly for economic 'co-benefits', and several State governments have followed with extensive transition plans.

In all three contexts, there is growing concern at the extensive impacts of climate change, with unprecedented floods and heatwaves. With this, there is a sharpening contradiction between prevailing policy and the climate change imperative, with significant disquiet at government climate policy especially in Australia and Germany. Public opinion on climate change appears to have converged. In India awareness can be low but among those who are aware of the issue concern is high (Lee et al., 2015; Thaker & Leiserowitz, 2014). In 2015 India almost topped the Pew Research Centre global poll on climate change, with 73% stating climate was their first concern (compared with just 34% in Germany and 37% in Australia; Pew Research, 2015, p. 18). By 2019 concern had risen across all three countries, with a strong majority stating climate change was affecting 'where I live', at 78% in Germany, 58% in Australia, and 70% in India (Pew Research, 2020). The key difference was in the perception of government inaction – in Germany and Australia almost two-thirds stated the government was 'doing too little'; in India, this fell to a little over a third.

The following discussion uses the three-part model to assess how these different national contexts enable or constrain climate movement organizations, what constituencies they are able to appeal-to and mobilize, and what alliances they are able to form.

Germany

The German Federal government has sought to be a leader in climate policy and has taken a major role in advocating for emission reduction efforts. Since the late 1990s, a series of laws and policy initiatives have encouraged the expansion of renewable energy and the decarbonization of the German economy under the banner of the *Energiewende* or 'German energy transition' (Fabra et al., 2015). The *Energiewende* frames decarbonization as a process of ecological modernization, 'transforming the trade-off between ecology and economy into a positive sum game (Hillebrand, 2013, p. 665). Since the 1980s support for renewable energy within the broader environmental movement in Germany has overlapped with opposition to nuclear power. The 2011 Fukushima nuclear accident prompted the German Federal government to pass new laws to accelerate the phase-out of nuclear power in Germany and to set ambitious targets for the decarbonization of the German economy and expansion of renewable energy (Renn & Marshall, 2016). Climate advocacy has its origins in the wider Green movement and is strongly linked to the anti-nuclear issue (Rucht & Roose, 2003). The German climate movement covers a broad political spectrum, from groups with an explicitly Marxist orientation to church organizations (Gentes et al., 2020, pp. 166–7).

Trade unions formally support climate action but have opposed the closure of coal mines and coal-fired power stations: this has changed in recent years for instance with the IGBCE, the union which represents coal miners, becoming involved in negotiating Germany's coal exit laws in 2019/2020. There are regular conflicts between groups over both fundamental aims and principles and forms of political action. Some organizations have highlighted the limitations of domestic climate policy, especially the continued commitment to fossil fuels, and the need to enable more distributed and socially-just forms of decarbonization (Greenpeace International, 2019). There is a forceful debate over transition models, with a strong and growing counterpoint from de-growth perspectives (Schmelzer, Vansintjan & Vetter, 2022).

At least until recently, 'advocacy for transition' has been the dominant model within the German climate movement, and is exemplified by the largest and most broadly based German climate advocacy organization, the Climate Alliance Germany (Klima Allianz Deutschland), which represents about 25 million members, and describes itself as a broad social alliance for climate protection, with 140 member organizations in the environmental, church, development, education, culture, health, and consumer protection sectors, and also youth organizations and unions. Since 2013 CAG has been affiliated with Climate Action Network Europe (an umbrella organization with more than 120 member organizations in 25 European countries). Broadly speaking, CAG fits within the 'advocacy for transition' framework: in its Statement of basic principles (Grundlagenpapier) it stresses that 'climate protection is a mainstream concern with a broad social basis', and that its capacity to mobilize many different social groups is due to its diversity of actors and the fact that it is above party politics (CAG, 2019). The Climate Manifesto (Das Klima-Manifest) advances arguments based on an ecological modernization approach, stressing that climate protection is a 'win-win' for all, which will create good jobs, democratize the energy supply and strengthen social participation (CAG, 2016). Fossil fuels must be left in the ground, and the burning of all fossil fuels must cease by 2050. The transition to a decarbonized economy and society is not positioned as a purely technocratic issue: 'social justice is not possible without climate protection, climate protection is not possible without social justice' (CAG, 2016), and the need for social transformation is stressed, albeit without any concrete blueprint for what such a transformation might entail.

From the beginning, there has been tension between broadly based organizations such as the Climate Alliance (Klima Allianz), which are aligned with mainstream climate politics and tend to assume a cooperative attitude to state institutions, and those with a 'system-critical' perspective shaped by the autonomous left and the social movements which emerged from the 1968 student protests (Kössler, 2013). Ende Gelände (EG; literally translatable as Stop the Site or Close the Ground) represents this latter tendency; it is a decentralized movement with an explicit focus on climate justice and direct action which clearly fits within the framework of 'advocacy for transformation'. EG undertook its first actions in Germany in 2015 to target fossil fuels at source, stating 'we have to take climate protection in(to) our own hands' (Ende Gelände, 2015). EG has a clear commitment to system change and has supported the 2020 'Program for Climate Justice Now', which notes the IPCC's call for 'rapid, far-reaching and unprecedented transformation in all areas of society' (IPCC, 2018) and argues for a shift away from 'unlimited growth based on unlimited resources'. The programme sets out a series of measures including a universal basic income, redistribution of wealth, energy democracy, 'recommunalisation' of electricity generation and grids, democratic reform of trade, and the international financial system (Ende Gelände, 2021). An 'Action Consensus' sets the framework for civil disobedience, stating that participants will 'blockade and occupy [fossil fuel] infrastructure with our bodies' but rejects any endangerment of people or destruction of property (Ende Gelände, 2021). Actions targeting coal mines across Germany have involved repeated clashes with police, civil disobedience, and arrests; recently the German Office for the Protection of the Constitution classified the Berlin branch of EG as a 'left extremist' group (Jansen, 2021).) dividing the parties in the Berlin Senate. Interestingly, a large number of German climate groups, including more mainstream organizations such as Bund (Friends of the Earth) and the youth wing of the Social Democrat party (SPD) responded by expressing their support for EG, stating that 'in light of the looming climate catastrophe we can fully understand that people feel compelled to undertake civil disobedience against coal mining' (Ende Gelände, 2020).

Taking a more 'emergency' perspective, the Fridays for Future campaign emerged in Germany in 2019 as part of the global climate strikes in 2019 inspired by Greta Thunberg, which mobilized a normative appeal centred on climate impacts (Buzogány & Scherhaufer, 2022). Their demands were tightly focused on achieving the goals of the Paris agreement, with net zero emissions by 2035, a complete exit from coal by 2030 (instead of the 2038 date foreseen by Germany's 'Coal Commission'), a 100% renewable energy supply by 2035, an immediate end to all fossil fuel subsidies, and a carbon tax on all greenhouse gas emissions (FfF, 2021). While the Climate Alliance Germany built a broad-based membership and mainstream political legitimacy, FfF galvanized public attention based on a normative appeal (Gentes et al., 2020; Haunss et al., 2020). Some unions such as the metalworkers' union (IG Metall) supported the Fridays for Future protests and encouraged members to participate. In mid-2019 German Chancellor Angela Merkel stated the FfF protests had accelerated the Federal government's decision-making in climate policy (Egenter & Wehrmann, 2019). Throughout FfFG maintained a politically-neutral appeal and positioned itself as a 'bipartisan movement' (Marquardt, 2020, p. 7). In 2020 FfG published a Wuppertal Institute study that called for 'prioritisation of climate protection across all sectors' including industry policy to retain economic competitiveness. On its website FfFG states that 'economic activity must not continue to exceed planetary boundaries', but neither here nor in the WI study does it demand system change or a move away from growth.

India

Until recently in India discontent over climate change has been externalized and managed by the post-colonial frame, preventing strong politicization (Sengupta, 2018). Reflecting the post-colonial legacy, climate change in India has largely been positioned as the responsibility of earlier industrializers, with advocacy groups differentiating 'luxury emissions' of the rich and 'survival emissions' of the poor (Agarwal & Narain, 1991). India played a dominant role in institutionalizing this discourse under the UNFCCC and the Kyoto Protocol which limited emissions-reduction responsibilities to developed countries. Since 2008 growing recognition of India's vulnerability to climate change has seen the country shift to support climate action that delivers co-benefits, for 'both development and climate gains' (Dubash, 2013, p. 191). Advocacy groups such as the Centre for Science and Environment played a central role in conceiving and framing this approach (Aamodt & Stensdal, 2017, p. 121). This important shift underpinned India's Paris commitments, including a pledge to source forty per cent of electricity from non-fossil fuel sources.

There is certainly a strong 'transition' focus, with emissions reduction and energy decarbonization presented as an important development opportunity. Many Indian climate advocacy groups have focused on energy systems and have adopted a largely apolitical focus on social innovation for 'sustainability' (Ylä-Anttila & Swarnakar, 2017). The resulting 'climate advocacy coalition' comprises a range of research-based policy advocacy NGOs and think-tanks (Aamodt & Stensdal, 2017). Many of these groups are centred on broad 'sustainable' transition agendas, for instance, to lobby for stronger climate commitments from the government and for domestic emission reduction measures (Swarnakar, 2019). Climate Action South Asia, based in New Delhi, has played a convening role in this broad 'transition' agenda. In 2021 it endorsed the 'Fossil Fuel Non-Proliferation Treaty' and in its policy paper *No Future for Fossil Fuels*called for a 'just transition' to address long-term community impacts by putting human and environmental well-being ahead of economic objectives' (CANSA, 2021, p. 29). The organization acts as a regional umbrella for international and national development NGOs and some environmental organizations, including 50 India-based organizations. It works closely with government agencies and is a policy advocate in national, regional, and international forums.

A second bloc of movement organizations focuses on fossil fuels rather than on the renewable transition, and clearly defines itself against the existing development model, pursuing more transformative agendas. Thaker and Leiserowitz found growing antagonism due to domestic vulnerability to climate change, which had 'led some to reassess India's development strategy' (2014, p. 12). Dubash et al. highlight the resulting pressure against fossil fuels and especially the coal industry (Dubash, 2019), while others emphasize a growing connection with long-running social justice struggles over land and livelihood (Bhushan et al., 2020; Swarnakar, 2019).

Here, there is an active process of channelling discontent from affected communities behind the wider climate justice agendas. Climate justice came very early to India, in 2002, with a large protest and counter-summit organized by the Delhi Conference of Parties of the UNFCCC. The mobilization was organized by the India Climate Justice Forum, which included a range of grassroots-based movement organizations under the National Alliance of Peoples Movements (Chatuverdi & Doyle, 2015). The Summit's Declaration addressed the unequal causes and impacts of climate change and targeted neoliberal climate policy, stating that 'market-based mechanisms and technological "fixes" currently being promoted by transnational corporations are false solutions and are exacerbating the problem'. Groups affirmed they should 'address the issue of climate change from a human rights, social justice and labour perspective', stating 'We reject the market-based principles that guide the current negotiations to solve the climate crisis: Our World is Not for Sale!' The 2002 Delhi Climate Justice Declaration was the inspiration for a wave of climate mobilization internationally: when the international Climate Justice Now! network was created in 2007 it included 29 India-based organizations (Swarnakar, 2019).

The climate justice approach offers an agenda that could dovetail with wider social and environmental justice struggles within India centred on contesting development projects. These are often led by local alliances of affected peoples and have a strong resource rights and livelihood focus. They come together in national contexts around the 2006 Forest Rights Act and in legal challenges (from 2010) through the National Green Tribunal. Long-running forest rights campaigns and climate concerns overlap in several states in central India where livelihood is threatened by coal mining (Goodman et al., 2020). Here, a new front has opened-up, for instance in the alliance between anti-coal campaigns in Singrauli and Greenpeace (Talukdar, 2016). There are also hard-fought struggles against coal-fired power stations, supported by national climate networks, that have successfully halted proposed power stations (Adve, 2020). There are also wider efforts at channelling discontent against coal at the government, which were quickly politicized when Greenpeace and other international NGOs were targeted as foreign agents undermining Indian development (Talukdar, 2018). The following public backlash demonstrated the extent to which climate campaigning was creating vulnerability for the political elite. The grassroots groups opposing fossil fuel projects were strengthened in 2019 with the formation of the 'South Asian Peoples Action on Climate Crisis' (SAPACC) in 2019 (Adve, 2020). At its founding meeting in Hyderabad, the group brought together key trade union confederations, forest rights groups, farmers and Adivasi organizations, and the Alliance of Peoples' Movements specifically to 'raise public awareness regarding climate change impacts and mobilize public action to mitigate it', creating a new national platform for a labour-climate justice alliance (SAPACC, 2021).

A third distinct wave of urban climate mobilization emerged in 2019 with a strong 'emergency' orientation. This was partly inspired by international initiatives, but also had local roots in the growing visibility of climate change impacts in India. During 2019, fourteen Extinction Rebellion (ER) groups were established. Fridays for the Future (FfFI) opened branches in more than sixty sites and were organizing monthly student strikes throughout the year (Adve, 2021). With this, there was a proliferation of local, urban-based groups. The 9-year old campaigner, Licyprita Kangujam, spoke at the 2019 COP and initiated urban campaigns through FfFI and 'The Child Movement', focusing on climate and smog (Sudeven, 2020). The Mumbai-based 'Let India Breathe' group also campaigned for climate action to address urban smog and, amongst many other actions across the country, led a petition calling for new planning regulations to address the anticipated inundation of Mumbai by 2050. The 'There Is No Earth B' group also has a strong focus on urban waste and is dedicated to 'inclusive, organic and decentralized climate action and advocacy'. These and other groups such as the youth-based Chennai Climate Action Group are engaged in on-the-ground campaigns and have attracted a large online following. A common feature of 'climate emergency' activism in India is its focus on the urban issues and contexts for mobilization, from smog impacts to tree conservation to urban development and seawater inundation. As the movement has gained momentum there has been a backlash from the state authorities. Groups that mobilized more than a million signatories against a watered-down Environmental Impact Act in 2020 had their websites blocked in 2020 (Joshi, 2020). In early 2021, several activists with FfFI and ER were arrested accused of spreading 'disaffection with the Indian state' after they circulated a call from Greta Thunberg to support farmers protesting against government agriculture policy (Ellis-Peteresen, 2021). There was a strong online campaign to secure their release but observers noted a chilling effect as young people sought to de-tag themselves from the emergent movement (Chatterjee, 2021). Overall though, these recent initiatives focus on politicizing impacts, reflecting a broad 'emergency' approach, and create a potential for a new urban front in Indian climate mobilization.

Australia

Australia has some of the world's largest reserves of fossil fuels. It is the fifth largest producer of coal, the largest exporter of metallurgical coal, and the second largest exporter of thermal coal. Along with Qatar, it is one of the two largest exporters of gas (Australian Government, 2021). National and State governments have resolutely supported fossil fuel expansion, marking the country as a climate pariah among high-income countries (Guardian/AAP, 2022). A wave of organizations centred on international climate negotiations emerged in the mid-2000s and soon refocused on mobilization and protest against coal and gas industries (Burgmann & Baer, 2012; Rosewarne et al., 2014). Fossil fuel corporates wield considerable influence in the country and political contention over climate policy has become a key factor in the rise and demise of party leaders and federal governments (Hudson, 2019; Wilkinson, 2020). Several State governments adopted ambitious climate action plans and there is a growing renewables sector, and an emerging renewables export sector (Ali et al., 2020; Garnaut, 2019). But resource-rich states, such as New South Wales, Queensland, and Western Australia, have also continued to support expanded fossil fuels, mainly for export, from which they derive royalty revenues. Federal fossil fuel subsidies have continued post-pandemic, at about AUD$10bln in 2021–2022 (Armistead et al., 2022). Climate advocacy organizations have sought to isolate the fossil fuel sector and advance an economy-friendly agenda for 'green' growth and jobs led by renewables (Ylä-Anttila et al., 2020). Climate litigation and divestment campaigns against fossil fuels have dovetailed with direct action against fossil fuel installations, with national climate action organizations and sector-based advocacy bodies, across health, agriculture, energy, and regional development, for instance. These have created new constituencies for climate action and helped politicize a widening range of policies in terms of climate change and energy transition. While efforts are often fragmented and marginalized, they have offered a means of sidestepping the Federal deadlock.

As with India, the climate movement in Australia has been centred on advocacy for transition. International Environment NGOs (ENGOs) had been engaged with climate questions since at least the late 1980s, for instance with the creation of the international Climate Action Network in 1988. Their counterparts in Australia, such as Greenpeace and the World Wildlife Fund established a series of campaigns for a stronger commitment from Australia in international negotiations. The 'Turn Down the Heat' campaign in 1997 for instance, led by the Australian Conservation Foundation, was targeted at the Kyoto COP negotiations. Some ENGOs such as State-based Conservational Councils helped convene a wave of locally-based 'Climate Action Groups' (CAGs), coordinated local actions in advance of the 2007 election and the 2009 COP. Initiatives such as the annual 'Walk Against Warming' from the mid-2000s were coordinated by ENGOs as part of the Australian Climate Action Network (CANA), which was established in 1992 mainly by environmental organizations. Through the mid-2000s a range of new climate-focused NGOs emerged, such as the Climate Institute in 2005, Beyond Zero Emissions and the Australian Youth Climate Coalition in 2006, Solar Citizens and the Climate and Health Alliance, both in 2010. All these groups joined with CANA, which continues to advocate for a range of 'transition' measures to enable emission reduction, including carbon pricing and emissions trading, albeit with a strong focus on energy transition and fossil fuel phase-out.

In contrast to CANA, a range of more transformative initiatives emerged from the early 2000s. These were inspired at least in part by the emergent climate justice movement. The international Rising Tide network was established in 2000 rejecting market-based measures and focusing on direct action against fossil fuel interests (its UK branch remained active in 2021). Soon after, an Australian Rising Tide was established in Newcastle, the world's largest coal port, which quickly became a focus for direct action against the coal industry. Mirroring mobilizations in the UK, Rising Tide helped convene a series of 'Climate Camps' from 2007, targeting coal infrastructure (Rosewarne et al., 2014). A national-level 'Climate Movement' conference was convened from 2007 to 2012 to coordinate collective actions. With the election in 2013 of a conservative government hostile to climate policy, climate justice campaigns shifted more explicitly into campaigns directed at new coal mines and coal seam gas (CSG).

With this, a new range of constituencies has been mobilized and a new wave of climate campaign organizations has followed. A large civil disobedience campaign of farmers against CSG was initiated by Lock the Gate in 2010; there was a new focus on supporting grassroots mobilizations against coal projects led by the Sunrise Project from 2012; the trade union movement has become directly engaged in a series of initiatives for 'just transition' in coal regions, such as the Hunter Jobs Alliance, and union-led groups such as Cooperative Power and Workers for Climate Action and have initiated campaigns for publicly-owned renewable energy; there have also been new divestment initiatives, for instance, Market Forces from 2013, and climate litigation initiatives such as

Environmental Justice Australia. Indigenous mobilizing on climate change emerged with SEED-Mob, created in 2014; the national 'Stop Adani' coalition was formed in 2016 to halt the emergence of a new coal region in the Galilee Basin in Queensland. These and many other NGOs, along with a range of local anti-coal groups have been convened under the 'Beyond Coal and Gas' banner at national gatherings held every two years since 2012.

Efforts at targeting the fossil fuel sector stand in contrast with the more 'emergency' centred frame for climate mobilization. This approach came to the fore in Australia in 2018 with the international appeal for urgent mobilization that came from Fridays for Future (FfFA) and Extinction Rebellion. The Australian Youth Climate Coalition played an important convening role, as did the 'Stop Adani' movement. Both had active branches across the country and an established infrastructure to facilitate the emergence of this new wave of mobilization. FfFA initiated school strikes in March 2018 and later developed into the 'School Strike Australia' campaign. Demands of the group at the September 2019 strike centred on a halt to new fossil fuels, 'including the Adani mine', along with a transition to 100% renewables domestically by 2030 including 'for exports', with 'just transition and job creation for all fossil-fuel workers and their communities' (FfFA, 2021). These were radical demands in the Australian context as they required the phase-out of coal, oil, and gas for export by 2030, as well as an immediate halt to new mines. The strike attracted over 300,000 people across the country, three times larger than the previous largest climate protest (at the Walk Against Warming in 2009). At the follow-up strike in September 2020, in the context of the COVID crisis, expectations were reduced, with placards symbolically replacing protestors at COVID-affected cities.

'Emergency' framing shifted into groupings such as Extinction Rebellion (ER) and Blockade Australia that retained a focus on disruptive action. ER was established in Australia in 2018, initially focused on Queensland and the proposed Galilee Basin coal region. From mid-2019 ER protests disrupted traffic in Brisbane, the State's capital city, including a 'rebellion day' protest in August with 380 arrests (Lynch, 2019; Smee, 2019). Its 'emergency' message gained wide support, with several hundred academics declaring they supported the ER demands (The Guardian, 2019); ER also drew public support from mainstream ENGOs, including the ACF and Greenpeace, in a significant realignment (Readfearn, 2019). Later actions focused on the fossil fuel sector with a blockade at a Melbourne mining conference in October 2019. The unprecedented wave of bushfires in late 2019 provoked mass 'climate emergency' mobilizations with tens of thousands joining a virtually spontaneous action under the banner 'Sydney is choking' in December 2019 (Zhou, 2019). The momentum was disrupted by the COVID crisis, but rolling protests continued to target coal, gas, and oil corporates across several capital cities in 2020 and into 2021. Many of these were heavily policed. When, from 2021, Blockade Australia staged a series of direct actions to 'build a political movement that can physically resist Australia's planet-destroying operations with disruptive and targeted action' , activists faced police harassment and pre-emptive arrest (Kilvert, 2022). ER meanwhile continued mounting symbolic actions, for instance painting the slogan 'duty of care' on the national parliament building after the announcement of the IPCC's sixth report in August 2021 (IPCC, 2021), echoing a Federal Court judgement that the Government was required under common law to consider the impacts of climate change for young people (ER, 2021, august 1).

Comparison and conclusions

The three types of climate movement discussed here are highly generative. They produce new ways of interpreting and acting on climate change, and in generating what Hulme called 'purposeful' climate agency (Hulme, 2010). The three approaches – transition, transformation, and emergency – produce different ways of addressing the crisis, in some ways conflicting, in some ways overlapping. The move from a science framing to a justice framing and most recently to a normative or moral framing arguably reflects a deepening of the climate movement as posing an 'emancipatory' challenge (Stuart et al., 2020). Research can play a role: the strategic leverage of any movement is contingent on context, but, as de Moore et al argue, there is an urgent need to understand 'whether and how the new climate activists will manage to make use of that' (Moor, et al., 2020, p. 624).

Across the three country cases, there are parallels in how leverage is produced for the movement. In all three countries climate movement organizations advancing agendas for transition are centred on socio-technical agendas for a decarbonized 'green economy' and on mobilizing public opinion behind scientific and policy expertise. Organizations aiming for more transformative agendas seek systemic change through decarbonization and take a more confrontational stance against fossil fuels, often in alliance with wider social forces. There are also organizations taking an 'emergency' approach, expressing outrage at the failure of climate policy and creating new constituencies, especially among young people, linked through place-based and intersectional solidarities. These outcomes are summarized in Table 1, below.

In each of the three countries, we can see that three distinct forms of power are mobilized – science and policy expertise, popular cross-society alliances, and a 'post-political' normative refusal. These have developed as parallel temporal developments reflecting the trajectory of climate concern, from a faith in anticipatory policy in the early 1990s, through to the assertion of justice and transformation in the face of policy failure from the mid-2000s, followed by the call for emergency action with unprecedented climate disruption from 2015.

The three sets of strategies politicize different sites and create different constituencies. In this way, they are read cumulatively, as a process issue-by-issue 'climatisation' (see Aykut, 2017). Climate policy debate has moved from fields of environmental policy into wider energy, industrial and financial spheres. International environmental NGOs have been displaced by a wave of climate-centred advocacy organizations in renewable energy, green industry, and ethical investment. Mobilization against fossil fuels, for 'just transitions' has brought new allies and constituencies, for instance in trade union and farmer organizations. A focus on the impacts of climate change has engaged affected peoples and created an inter-generational impetus, securing future generations for climate action.

Arguably the three approaches dovetail and may be complementary, for instance with transition reformers benefiting from the agenda-setting actions of frontline campaigners. Beyond tactical complementarity, it may be possible to envisage a more strategic shared agenda, centred on the

Table 1. Summary of the argument: three types of climate movement.

	Movement goals	Mode of action	Targets	Organization
Transition	Climate policy for socio-technical policy solutions and market measures	Expert advocacy for 'green economy', sustainability and renewables	Institutional policy-making, UN, national, sub-national	NGO advocacy, think-tanks, transnational advocates
Transformation	Fossil fuel sector shut-down, decarbonized social alternatives and sufficiency or de-growth	Grassroots cross-sector mobilization; cross-sector alliances for social veto and transformation	Fossil fuel installations, local-national alliances, prefigurative projects	Activist NGO organizing, movement campaigns
Emergency	Politicizing the impacts of climate change, now and into the future	Mobilization of normative outrage at policy failure, intersectoral, intergenerational	Urban-based symbolic disruption, policy-centred demands	Social media, network politics, place-centred intersectional action

issue of societal 'development' under climate change, in a climatized 'movement society'. Climate change politicizes development hierarchies, both domestically and internationally, and offers a 'multidimensional critique of [the] industrialization and growth paradigm' (Lele, 2012, p. 215). Arguably, with growing climate change impacts across North and South, this offers an overarching agenda. Climate change is forcing a reckoning with dominant ideologies of developmentalism, and the industrial growth paradigm, posing new questions for the underlying social model.

Clearly, there are tensions. Efforts at building a widescale popular movement for transformation are directly undermined by neoliberal climate policy measures that create new injustices. Likewise, authorities routinely create divisions in the movement by criminalizing the more disruptive climate actions. This raises the broader question of climate democracy and democratic participation, where people are able to develop shared agendas for transformation. Perhaps the clearest evidence of this is in India, as exemplified by the grassroots struggles against coal mining by forest rights groups, farmers, and Adivasi organizations. In high-income, fossil-fuel dependent economies, such as Australia and Germany, climate justice centres the costs of transition and the impact of decarbonization on affected workers and regions. But here, too there is potential for new alliances and modes of democratic participation, especially in regions most affected by the transition. There is also the possibility that the increasingly widespread social experience of climate impacts and extreme weather events, such as the 2019–2020 bushfires in Australia, 2021 floods in Germany, and workers' exposure to extreme heat in India in 2022, will create both greater urgency and political impetus for transformative climate mobilization. The field remains highly dynamic and generative, signalling the vitality of what has become a sustained climate movement.

Disclosure statement

No potential conflict of interest was reported by the authors.

Funding

This work was supported by Australian Research Council: [Grant Number DP180103868 Decarbonising Electricity].

ORCID

James Goodman http://orcid.org/0000-0003-3211-3416
Tom Morton http://orcid.org/0000-0002-4255-2992

References

Aamodt, S., & Stensdal, I. (2017). Seizing policy windows: Policy influence of climate advocacy coalitions in Brazil, China, and India, 2000–2015. *Global Environmental Change*, *46*, 114–125. https://doi.org/10.1016/j.gloenvcha.2017.08.006

Adve, N. (2020, February 26). South Asian coalition links climate with social struggles. News Blog, *The Ecologist*.

Adve, N. (2021, April 2). Coming of age of India's youth climate movement. *India Forum*.

Agarwal, A., & Narain, S. (1991). *Global warming in an unequal world*. Centre for Science and Environment.

Ali, S., Svobodova, K., Everingham, J. A., & Altingoz, M. (2020). Climate policy paralysis in Australia: Energy security, Energy Poverty and Jobs. *Energies*, *13*(18), 4894. https://doi.org/10.3390/en13184894

Anshelm, J., & Hultman, H. (2015). *Discourses of global climate change: Apocalyptic framing and political antagonisms*. Routledge.
Aranof, K., Battistoni, A., Cohen, D. A., & Riofrancos, T. (2019). *A planet to Win: Why We need a green New deal*. Verso.
Armistead, A., Campbell, R., Littleton, E., & Parrott, S. (2022). *Fossil fuel subsidies in Australia*. The Australia Institute.
Australian Government. (2021). *Australia's energy commodity resources: gas*. Geosciences Australia.
Aykut, S., Foyer, J., & Morena, E. (Eds.). (2017). *Globalising the climate. COP21 and the climatisation of global debates*. Routledge.
BBC. (2021, November 5). COP26: Greta Thunberg tells protest that COP26 has been a 'failure'. *BBC News*.
Beck, U. (2015). Emancipatory catastrophism: What does it mean to climate change and risk society? *Current Sociology*, *63*(1), 75–88. https://doi.org/10.1177/0011392114559951
Bellamy-Foster, J. (2002). *Ecology against capitalism*. Monthly Review Press.
Bellamy-Foster, J. (2019). On fire this time. *Monthly Review*, *17*(6), 1–17. https://doi.org/10.14452/MR-071-06-2019-10_1.
Bennholdt-Thomsen, V., & Mies, M. (1999). *The subsistence perspective*. Zed.
Berglund, O., & Schmidt, D. (2020). *Extinction rebellion and climate change activism*. Palgrave-Springer.
Bhushan, C., Banerjee, S., & Agarwal, S. (2020). *Just transition in India*. Sustainability Innovations.
Brick, P., & Cawley, M. (2008). Producing political climate change: The hidden life of US environmentalism. *Environmental Politics*, *17*(2), 200–218. https://doi.org/10.1080/09644010801936107
Bullard, N., & Müller, T. (2012). Beyond the 'Green Economy': System change, not climate change? *Development*, *55*(1), 54–62. https://doi.org/10.1057/dev.2011.100
Burgmann, V., & Baer, H. (2012). *Climate politics and the climate movement in Australia*. Melbourne University Press.
Buzogány, A., & Scherhaufer, P. (2022). Framing different energy futures? Comparing Fridays for future and extinction rebellion in Germany. *Futures*, *132*, 102904. https://doi.org/10.1016/j.futures.2022.102904
Cabello, J., & Gilbertson T. (Eds.). (2015). *Paths beyond Paris: Movements, Action and Solidarity towards Climate Justice*. Carbon Trade Watch.
CAG. (2016). *Klima-Manifest der klima allianz deutschland*.
CAG. (2019). *Grundlagenpapier der Klima Allianz Deutschland, adopted in plenary session 18.09.2019 in Berlin*.
CANSA. (2021). *No future for fossil fuels*. Climate Action Network South Asia.
The Guardian. (2022, March 22). Chattering classes': Australian government dismisses UN secretary general's climate criticism. *The Guardian*.
Chatterjee, S. (2021, February 15). Fridays for Future was on govt radar long before Disha Ravi's arrest: An inside view. *The News Minute*.
Chatterton, P., Featherstone, D., & Routledge, P. (2012). Articulating climate justice in Copenhagen. *Antipode*, *45*(3), 602–620. https://doi.org/10.1111/j.1467-8330.2012.01025.x
Chatuverdi, S., & Doyle, T. (2015). *Climate terror: A critical geopolitics of climate change*. Springer Nature.
Dahlin, J., & Fredriksson, M. (2017). Extracting the commons. *Cultural Studies*, *31*(2-3), 253–276. https://doi.org/10.1080/09502386.2017.1303428
D'Alisa, G., Demaria, F., & Kallis, G. (2010). *Degrowth: A vocabulary for a new era*. Routledge.
Davis, M. (2010). Who will build the Ark? *New Left Review*, *61*, 29–46.
Della Porta, D., & Parks, L. (2014). Framing processes in the climate movement. In M. Dietz, & H. Garrelts (Eds.), *Handbook of the climate change movement* (pp. 1–12). Routledge.
de Moor, J., De Vydt, M., Uba, K., & Wahlström, M. (2020). New kids on the block: Taking stock of the recent cycle of climate activism. *Social Movement Studies*, *20*(5), 619–625. https://doi.org/10.1080/14742837.2020.1836617
De Young, R., & Princen, T. (2012). *The localization reader: Adapting to the coming downshift*. MIT Press.
Di Chiro, G. (2008). Living environmentalisms: Coalition politics, social reproduction, and environmental justice. *Environmental Politics*, *17*(2), 276–298. https://doi.org/10.1080/m09644010801936230
Dietz, M., & Garrelts, H. (2014). *Routledge handbook of the climate change movement*. Routledge.
Dubash, N. (2013). The politics of climate change in India: Narratives of equity and cobenefits. *Wiley Climate Change*, *4*(3), 191–201. https://doi.org/10.1002/wcc.210

Dubash, N. (2019). *India in a warming world*. Oxford University Press.
Egenter, S., & Wehrmann, B. (2019, July 19). Merkel backs national CO_2 price, cites Greta-effect on policy. *Clean Energy Wire*.
Ellis-Peteresen, H. (2021, February 15). India: Activist arrested over protest 'toolkit' shared by Greta Thunberg. *The Guardian*.
Ende Gelände. (2015). *Keep it in the Gorund! Action conference Leipzig*. Ende Gelände.
Ende Gelände. (2020). Solidaritätserklärung der NGOs mit Ende Gelände nach Angriff des Verfassungsschutz, 22 June, Ende Gelände.
Ende Gelände. (2021). *Aktionskonsens and sofortprogramm klimagerechtigkeit*.
ER. (2021, August 15). We have a Duty of care to Reclaim our future. *Extinction Rebellion*.
Fabra, N., Matthes, F., Newbery, D., Colombier, M., Mathieu, M., & Rüdinger, R. (2015). *The energy transition in Europe: Initial Lessons from Germany, the UK and France: Towards a low carbon European power sector*. Centre on Regulation in Europe (CERRE).
FfF. (2021). Unsere Forderungen an die Politik. Fridays for the Future Germany.
Foust, C., & Murphy, W. (2009). Revealing and reframing apocalyptic tragedy in global warming discourse. *Environmental Communication*, *3*(2), 151–167. https://doi.org/10.1080/17524030902916624.
Garnaut, R. (2019). *Superpower: Australia's low-carbon opportunity*. Latrobe University Press.
Gentes, T., Löning, L., & Trapp, A. (2020). Nähe und distanz, Das Verhältnis zwischen Umwelt-NGOs und Fridays for future. In S. Haunss, & M. Sommer (Eds.), *Fridays for future. Die Jugend gegen den Klimawandel. Konturen der weltweiten Protestbewegung transcript* (pp. 163–180). Verlag Bielefeld.
Giacomini, T., & Turner, T. (2015). The 2014 people's climate March and flood Wall Street civil disobedience. *Capitalism, Nature, Socialism*, *26*(2), 27–45. https://doi.org/10.1080/10455752.2014.1002804
Goodman, J., Connor, L., Ghosh, D., Kohli, K., Marshall, J. P., Menon, M., Mueller, K., Rosewarne, S., Morton, T., & Pearse, R. (2020). *Beyond the coal rush: A turning point for global energy and climate policy?* Cambridge University Press.
Goodman, J., & Salleh, A. (2013). The 'Green Economy': Class hegemony and counterhegemony. *Globalizations*, *10*(3), 411–424. https://doi.org/10.1080/14747731.2013.787770
Goodman, K. (2022). *Social movements and climate change: "Climatizing" society from within. Climate Science, Oxford Research Encyclopedia*. Oxford University Press.
Greenpeace International. (2019, January 26). Greenpeace reaction to Germany's plan to phase-out coal, urges haste. *Press Release*.
Hadden, J. (2015). *Networks in contention: The divisive politics of climate change*. Cambridge University Press.
Hales, R., & Mackey, M. (2021, November 15). The ultimate guide to why the COP26 summit ended in failure and disappointment (despite a few bright spots). *The Conversation*.
Hampton, P. (2015). *Workers and trade unions for climate solidarity*. Routledge.
Harding, J. (2020). The arrestables. *London Review of Books*, *42*(8).
Harvey, D. (1996). *Justice, nature and the politics of difference*. Blackwell.
Haunss, S., Rucht, D., Sommer, M., & Zajak, S. (2020). Germany. In M. Wahlström, M. Sommer, P. Kocyba, & M. De Vydt (Eds.), *Protest for a future: Composition, mobilization and motives of the participants in Fridays For Future climate protests on 15 March, 2019 in 13 European cities* (pp. 69–81). Institute for Protest.
Hillebrand, R. (2013). Climate protection, energy security, and Germany's policy of ecological modernization. *Environmental Politics*, *22*(4), 664-682. https://doi.org/10.1080/09644016.2013.806627
Hudson, M.. (2015, November 6). Citizens, arrests, and 7-metre dinosaurs: the history of UN climate summit protests. *The Conversation*.
Hudson, M. (2019). 'A form of madness': Australian climate and energy policies 2009–2018. *Environmental Politics*, *28*(3), 583–589. https://doi.org/10.1080/09644016.2019.1573522
Hulme, M. (2010). Mapping climate knowledge: An editorial essay. *Wiley Interdisciplinary Reviews: Climate Change*, *1*(1), 1–7. https://doi.org/10.1002/wcc.3
Illius, S. (2021, November 13). COP26: Global south 'feels cheated' again. *The Business Standard*.
IPCC. (2018). *Global warming of 1.5°C: Summary for policymakers*. Intergovernmental Panel on Climate Change.

IPCC. (2021). *Climate change 2021: The physical science basis: Summary for policy-makers.* Intergovernmental Panel on Climate Change.

Jamison, A. (2010). Climate change knowledge and social movement theory. *Wiley Interdisciplinary Reviews: Climate Change, 1*(6), 811–823. https://doi.org/10.1002/wcc.88

Jansen, J. (2021, May 26). Linkspartei und SPD in Berlin streiten über linksradikale Klimaschützer. *Der Spiegel.*

Joshi, S. (2020, July 15). India wants to let polluters self-report pollution. Now, groups against this stupid plan have had their sites blocked. *Vice India.*

Kilvert, N. (2022, July 22). Who are Blockade Australia, and why are some members facing charges that carry up to 10 years in prison? *ABC Science.*Who are Blockade Australia, and why are some members facing charges that carry up to 10 years in prison? *ABC Science.*

Kössler, G. (2013). Die klimabewegung in deutschland. In M. Dietz, & H. Garrelts (Eds.), *Die internationale kliambewegung: Ein handbuch* (pp. 185–204). Wiesbaden Springer S.

Kovel, J. (2007). *The enemy of nature: The end of capitalism or the end of the world.* Zed.

Lee, T., Markowitz, E., Howe, P., Ko, C., & Leiserowitz, A. (2015). Predictors of public climate change awareness and risk perception around the world. *Nature Climate Change, 5*(11), 1014–1020. https://doi.org/10.1038/nclimate2728

Lele, S. (2012). Climate change and the Indian environmental movement. In N. Dubash (Ed.), *Handbook of climate change and India: Development, politics and governance* (pp. 208–215). Oxford University Press.

Lipshutz, R., & McKendry, C. (2011). Social movements and global civil society. In J. Dryzek, R. B. Norgaard, & D. Schlosberg (Eds.), *The Oxford handbook of climate change and society* (pp. 369–383). Oxford University Press.

Luke, T. (2014). The climate imaginary. *Current Sociology, 63*(2), 280–296. https://doi.org/10.1177/0011392114556593

Lynch, L. (2019, December 16). Police use new powers against climate activists just once in a year. *Sydney Morning Herald.*

Marquardt, J. (2020). Fridays for future's disruptive potential: An inconvenient youth between moderate and radical ideas. *Frontiers in Communication, 5*(48), 1–18. https://doi.org/10.3389/fcomm.2020.00048

Martiskainen, M., Axon, S., Sovacool, B. K., Sareen, S., Furszyfer Del Rio, D., & Axon, K. (2020). Contextualizing climate justice activism: Knowledge, emotions, motivations, and actions among climate strikers in six cities. *Global Environmental Change, 65*, 102180. https://doi.org/10.1016/j.gloenvcha.2020.102180

Moll, A., & Sonnenfeld D. (Eds.). (2000). *Ecological modernisation around the world perspectives and critical debates.* Routledge.

Moor, J., Vydt, M., Uba, K., & Wahlstrom, M. (2020). New kids on the block: Taking stock of the recent cycle of climate activism. *Social Movement Studies, 20*(5), 619–662. https://doi.org/10.1080/14742837.2020.1836617

Mueller, T., & Passakis, A. (2009). Green capitalism and the climate. In Brand U., Bullard N., Lander E., & Mueller T (Eds.), *Contours of climate justice* (pp. 54–61). Dag Hammarskjold Foundation.

North, P. (2011). The politics of climate activism in the UK: A social movement analysis. *Environment and Planning, 43*(7), 1581–1598. https://doi.org/10.1068/a43534

Nulman, E. (2015). *Climate change and social movements.* Palgrave Macmillan.

Pew Research. (2015). *Global concern about climate change: Broad support for limiting emissions. Report.* Pew Research Center.

Pew Research. (2020). *Science and scientists held in high esteem across global publics. Report.* Pew Research Center.

Readfearn, G. (2019, October 9). Extinction Rebellion risks polarising Australian public on climate, veteran activist says. *The Guardian.*

Reitan, R., & Gibson, S. (2012). Climate change or social change? *Globalizations, 9*(3), 395–410. https://doi.org/10.1080/14747731.2012.680735

Renn, O., & Marshall, J. P. (2016). Coal, nuclear and renewable energy policies in Germany: From the 1950s to the "Energiewende". *Energy Policy, 99*, 224–232. https://doi.org/10.1016/j.enpol.2016.05.004

Reuters. (2021, November 14). *Deeply sorry': UK's Sharma offers apology for last-minute changes to climate deal.* Reuters.

Rosewarne, S., Goodman, J., & Pearse, R. (2014). *Climate upsurge: The ethnography of climate movement.* Routledge.

Rucht, D., & Roose, J. (2003). Germany. In C. Rootes (Ed.), *Environmental protest in Western Europe* (pp. 80–108). Oxford University Press.

SAPACC. (2021). Keep the climate, change the economy. South Asian people's action on climate crisis launch meeting, 18–21 September, Hyderabad.

Saunders, C., Doherty, B., & Hayes, G. (2020). *A New climate movement? Extinction rebellion's activists in profile.* CUSP Working Paper 25.

Schmelzer, M., Vansintjan, A., & Vetter, A. (2022). *The future Is degrowth: A guide to a world beyond capitalism.* Verso.

Sengupta, S. (2018). The climate change discourse in India. Is a social movement possible? In Savyasaachi (Ed.), *Intractable conflicts in contemporary India narratives and social movements* (pp. 219–233). Routledge.

Smee, B. (2019, August 11). Extinction Rebellion: Hitting a nerve at Australia's climate flashpoint. *The Guardian.*

Spaiser, V., Nisbett, N., & Stefan, C. (2022). "How dare you?" The normative challenge posed by Fridays for Future. *PLOS Climate, 1*(10), https://doi.org/10.1371/journal.pclm.0000053

Stuart, D., Gunbderson, R., & Petersen, B. (2020). The climate crisis as a catalyst for emancipatory transformation: An examination of the possible. *International Sociology, 35*(4), 433–456. https://doi.org/10.1177/0268580920915067

Sudeven, P. (2020, November 30). India should make climate education compulsory: Nine-year-old activist Licypriya Kangujam. *The Hindu.*

Swarnakar, P. (2019). Climate change, civil society, and social movements. In N. Dubash (Ed.), *India in a warming world* (pp. 253–272). Oxford University Press.

Swyngedouw, E. (2010). Apocalypse forever? Post-political populism and the spectre of climate change. *Theory, Culture, Society, 27*(2/3), 212–232. https://doi.org/10.1177/0263276409358728

Talukdar, R. (2016). Hiding neoliberal coal behind the Indian poor. *Journal of Australian Political Economy, 78*, 132–158.

Talukdar, R. (2018). Sparking a debate on coal: Case study on the Indian Government's crackdown on Greenpeace. *Cosmopolitan Civil Societies: An Interdisciplinary Journal, 10*(1), 46–60. https://doi.org/10.5130/ccs.v10i1.5602

Tarrow, S. (2011). *Power in movement: Social movements and contentious politics* (Third Edition). Cambridge University Press.

Thaker, J., & Leiserowitz, A. (2014). Shifting discourses of climate change in India. *Climatic Change, 123*(2), 107–119. https://doi.org/10.1007/s10584-014-1059-6

The Guardian. (2019, September 20). We declare our support for Extinction Rebellion': An open letter from Australia's academics. *The Guardian.*

Tokar, B., & Gilbertson, T. (Eds.). (2020). *Climate justice and community renewal: Resistance and grassroots solutions.* Routledge.

Wallerstein, I. (2011). Structural crisis in the world-system: Where do we go from here? *Monthly Review, 62*(10), 31–39. https://doi.org/10.14452/MR-062-10-2011-03_2

Wilkinson, M. (2020). *The carbon club.* Allen & Unwin.

Ylä-Anttila, T., Gronow, A., Karimo, A., Goodman, J., & da Rimini, F. (2020). Breaking the treadmill? Climate change policy networks and the prospects for Low carbon futures in Australia and finland. *Society & Natural Resources, 33*(11), 1380–1398. https://doi.org/10.1080/08941920.2020.1802540

Ylä-Anttila, T., & Swarnakar, P. (2017). Crowding-in: How Indian civil society organizations began mobilizing around climate change. *The British Journal of Sociology, 68*(2), 273–292. https://doi.org/10.1111/1468-4446.12251

Zhou, N. (2019, December 11). Sydney climate protest: Thousands rally against inaction amid bushfire and air quality crisis. *The Guardian.*

Climate, violence, resource extraction and ecological debt: global implications of an assassination on South Africa's coal mining belt

Patrick Bond

ABSTRACT

Extractivism has attracted inspiring resistance in South Africa, but far-reaching lessons from one site of struggle in KwaZulu-Natal Province are sobering. The October 2020 assassination of Fikile Ntshangase, an anti-coal activist, reflects difficult political-economic and political-ecological terrain. The critical role of gendered activism in former apartheid-era Bantustans underlay concrete problems Ntshangase faced when confronted by male coal-mine labourers. There was no 'Just Transition' programme to decarbonize the mine at the time. The corporation that benefited from the assassination, Petmin, was expanding into Ntshangase's village. In the process, not only did the need to cease coal mining become of enormous global concern, but Petmin's role in international circuits of capital also gave rise to a new round of global-local and socio-ecological linkages. Some involve the World Bank while others entail solidarity with victims of the same firm near the United States city of Cleveland.

Fikile Ntshangase died aged 63 on 22 October 2020. She was assassinated by three men who entered her house – with several grandchildren playing nearby – and shot her six times, following her success in stalling a local mine's expansion. The murder, still unsolved in late-2022, was just one reflection of the multiple forms of fossil-fuel threats in the Somkhele region of northern KwaZulu-Natal, on the border of Africa's oldest proclaimed nature reserve, Hluhluwe-iMfolozi. The beneficiary of the killing was, at first blush, the vast Tendele coal mine complex, owned by Johannesburg-based (white-dominated) Petmin and originally financed in part by the World Bank via a venture capital fund meant to help upcoming black capitalists. Reparations demands on Petmin and investors by Ntshangase's lawyers (Tembeka Ngcukaitobi and Kirsten Youens) were advanced in court ten days after the assassination.

Hence the unity of environmental, community and potentially even labour movements at these sites of struggle – driven increasingly by militant youth – must in future be the renewed objective of activists and their allies, so that memories of what Ntshangase and her community have contributed never fade, but instead refresh and reverberate. Their lessons are inspiring but also sobering, in part because of the various forms of violence meted out in northern KwaZulu-Natal, and the opportunity presented by 'Just Energy Transition' financing to start the reparations process.

Generous state permission is still given to South Africa's mining houses, to dispossess small farmers of land and water, in Petmin's case so as to dig out and wash more coal, no matter how drought- and pollution-creating this form of energy has been. Coal mining and combustion in eastern South Africa generates several of the world's worst air-pollution hotspots for nitrous oxide and SO2, as Greenpeace has demonstrated, not to mention the water despoliation litigated regularly by the Centre for Environmental Rights (2020).

In describing the processes associated with contestation of coal mining in these circumstances, it has been useful for this author to maintain close connections with local activists and their lawyers. This occurs in search of a praxis-based epistemology in which new knowledge about the interrelationships emerges from observing power struggles between society, state and capital, global to local. The implications of this methodology include not only describing far-reaching connectivities (usually associated with corporate campaign research) that link principles of anti-extractivism and analysis of power balances, but also a sober assessment of the strategies, tactics and alliances that have emerged, whether they work or not, and what kinds of internationalist politics remain to be fully explored.

An exploration of the strengths and weaknesses of oppositional climate politics is one way to pay tribute to Ntshangase. The period after her assassination was difficult, as environmental-justice groups struggled to move forward when recognizing not only the urgent need to defend against immediate enemies – in her case, the coal firm Tendele – but also the potentials for *uniting with labour* to demand and win a genuine Just Transition. On the one hand, the two largest unions in the country – metalworkers and mineworkers – shifted radically to a pro-coal standpoint. On the other, the Just Transition envisaged by environmental justice activists logically emphasized detoxing poisoned land and water, while adding a food sovereignty strategy that often emanates from rural women's campaigning against both corporate agriculture and ethnic patriarchal power (Bond et al., 2021). Likewise, as shown below, there are many *local oppositional motivations for anti-extractivism,* but ultimately the approach activists critical of coal mining take may have to resort more to broader *international alliances associated with the climate crisis.* The activists and lawyers have long stressed Tendele's failure to consult affected communities, but at some stage there is likely to be a realization that such communities now stretch to the global scale given the damage done by thermal coal's CO2 emissions.

Fighting coal extraction and ecological destruction

Somkhele is an extended village bordering of Hluhluwe-iMfolozi Park (HIP), a vital site of conservation. The nature reserve was established in 1895 and became most famous for reviving the near-extinct Southern White Rhinoceros in the 1960s. But starting in 2006, Petmin's Tendele coal mine began destroying the land farmed – and river water consumed – by Ntshangase's allies in MCEJO, which at peak had 4000 members.

In March 2016, after a long period without rain, *Business Day* newspaper interviewed Petmin executive Bradley Doig (2016): 'You need water to wash the coal and if the drought persists, what are the next six months going to be like? Will you have to scale back production quite substantially?' Doig answered:

> We had lots of contingency plans in place obviously and one of them was we found a lot of underground water in the vicinity of our mine so we're in a very fortunate position right now that we have sufficient water available to run all three plants at maximum capacity to the end of this calendar year.

Surprised, the interviewer asked, 'Is this something that you should do?' Doig's (2016) answer: 'Yes, we believe it's necessary. The market is there for our product, we have the water available, and obviously we're still anticipating rainfall. But assuming it does not rain at all, we still have sufficient water for 12 months'.

But some nearby villages certainly did not have sufficient clean water. Three years later, 29 MCEJO women were arrested in a protest against Tendele's water grabbing and state water-supply failure (South African Food Sovereignty Campaign, 2021). Then in April 2020, the Covid-19 lockdown reinforced the extreme urgency of improved hygienic infrastructure. As Durban's main newspaper reported,

> The desperate search for clean, free water in the cluster of villages that make up Somkhele has long been a women's burden: walking for as much as six hours, risking sexual harassment and rape. And now, with the threat of Covid-19, the struggle has become more harrowing for the women of Somkhele. (Bega, 2020)

The main lawyer for Ntshangase and MCEJO, Kirsten Youens of the public interest firm AllRise, demanded the water minister improve access to tankers, boreholes and taps:

> As you may be aware, these communities are living on the border of the Somkhele open cast coal mine, which causes pollution of the rivers and streams as well as of the run-off rainwater that the communities collect from their roofs into JoJo tanks. This makes it impossible for people to utilize ways in which they used for water security ... Those without access to clean, reliable water are immediately at risk. Our client and the communities living in these areas are in need of urgent government intervention for the protection and fulfilment of their basic Constitutional rights, not only to address the current Covid-19 crisis but also to provide sustainable access to water. (Bega, 2020)

Ntshangase's close allies in an Africa-wide anti-extractivist feminist network, WoMin, also complained,

> With the onset of the Covid-19 pandemic, these communities are facing a major crisis. Last week, there was no water, not even the local clinic has access to water ... The only water tanks available to communities are the ones placed by the municipality in central locations, but as of now, they are empty. The Umfolozi River and dam were fenced off so that the community could not have access to these water sources. The little water remaining has dried up due to a series of droughts. (Bega, 2020)

The core problem remained the Tendele coal mine. A few months later, Fikile Ntshangase told friends she was approached by mine management who offered a \$24,000 bribe to support even more coal digging, in spite of the terrible damage it was causing (Walters, 2021). Worsening intimidation by the pro-mining lobby, including the provincial government, is documented in a chilling statement by a group of allied organizations and lawyers (Mfolozi Community Environmental Justice Organization, 2020). But Ntshangase refused to withdraw her court challenge to the expansion project: 'I refused to sign. I cannot sell out my people. And if need be, I will die for my people' (Walters, 2021).

This refusal was widely considered the motivation for Ntshangase's murder. She was reportedly one of several names on a local hit list (Pitt, 2020). The mine's owners argued, in rebuttal (while denying involvement in the hit job) (Kockett & Hattingh, 2020), thatnot only were several hundred local jobs at stake. There was also a 20 percent community-worker shareholding in Tendele mine, a share that had to be purchased through dividend payments (and hence delivered no benefits in the short term to the community). That share was mainly controlled by the Mpukunyoni Traditional Council and three trade unions. Tendele's white-male executives had used local financier Nedbank's funding in 2015 to buy them off not only with a 16 percent 'community' share, but an additional 4 percent for workers represented by three unions, including the one best known for fighting Lonmin at Marikana: the Association of Mineworkers and Construction Union (AMCU).

This was especially tragic, because the other two were not a surprise: the National Union of Mineworkers, founded and led by future South African President Cyril Ramaphosa during the 1980s and still allied with the ruling party; and 'Solidarity', a mainly white-Afrikaner union whose allied 'cultural' policy lobbyists – the apartheid-minimizing 'AfriForum' (Modjadji & Goba, 2018) – impressed Donald Trump with bogus claims of mass murder of white farmers (Msimang, 2018). These unions' traditional focus is on increasing jobs and pay in the mining sector, without much attention to the broader 'resource-cursing' so evident in Somkhele (Cock, 2019).

Without a Just Transition as an alternative, the mineworkers' unions in the coal fields are logically opposed to Ntshangase and the broader environmental – especially climate – considerations. But even AMCU's national treasurer (Jimmy Gama) described Petmin's 'Economic Benefits Sharing' scheme (based on an extremely expensive borrow-to-buy strategy) as 'an innovative Broad-Based Empowerment approach that will effect real economic benefits sharing to bridge the gap between social and mining economics' (Petmin, 2015). And in 2022, this standpoint was amplified in a paper delivered by leadership of the National Union of Metalworkers of South Africa (2022) to its highest decision-making body:

> While the government has committed itself to move away from coal, we notice that coal exports from South Africa, and indeed globally, have skyrocketed. Many countries in the world, especially those that are in the BRICS axis, refused to make such commitments. China continues to build coal-fired power stations, with clean-coal technologies. The fundamental question is: why is South Africa not doing the same given the abundance of coal? … we should be wary to be made to move away from our strengths, such as coal, when we are ill-prepared, without even discussions on the possibilities of deploying clean-coal technologies.

Aside from labour hostility to ceasing coal extraction, there was an additional 'conservation NGO' that favoured Tendele mine's expansion, albeit not westward towards the animal-rich HIP but instead north-easterly *into low-income residential areas* (including Ntshangase's village): the white-dominated Wildlife and Environment Society of South Africa (2014). The group expressed satisfaction that when Petmin:

> discovered another viable coal seam close to the HIP boundary a few years ago, they acknowledged the tangible value of the iMfolozi wilderness area and chose not to mine this seam; instead they laudably [*sic*] selected a seam further away to reduce the negative impacts on the HIP.

In contrast, several trusted solidarity groups with an environmental justice philosophy – groundWork, Earthlife Africa, Mining Affected Communities United in Action, Women Against Mining United in Action, ActionAid, Southern Africa Human Rights Defenders Network, Global Environmental Trust and WoMin – have provided support to MCEJO (2020), alongside the group's lawyers from AllRise Attorneys and the Centre for Environmental Rights.

To be sure, the conservationist agenda is not always anti-people in this region, though KwaZulu-Natal Province has witnessed more than a century and a quarter of forced removals of Zulu peasants to make way for nature reserves (mainly enjoyed by whites), whether public or privately owned. A different model was applied by HIP's most famous employee, Ian Player, who is credited with raising the park's white rhino population (then unique in the world) from 440 in 1958 to what is now a global census of 18,000. This was possible because Player – who was the brother of the infamous apartheid golf ambassador Gary Player – worked closely with border communities, especially through legendary Somkhele guide Magqubu Ntombela (Berry, 2015).

Even before Player died in 2014, Tendele mine's blasting for coal and air pollution had become a major threat to animal welfare, at a site the white rhino is again threatened, this time by poachers. And if the state permits a future Tendele dig to directly border HIP, that will also intensify poacher penetration, according to the Global Environmental Trust (Save our Wilderness, 2021). In the calendar year of 2019, before the Covid-19 lockdown reduced the carnage, there were 126 rhinos killed for their valuable horns in HIP, mainly sold by Chinese and Vietnamese crime syndicates, though lockdown meant the number fell to 88 in 2020 (Ngema, 2020). This is, in short, an ideal site for conservationists to unite with communities and climate activists to fight coal. Environmental justice is core to this campaign, as is eco-feminism, which forms a strong current in local and national politics (Madhanagopal et al., 2022). In 2017, WoMin (2017) documented the gendered eco-social atrocities caused by Tendele coal mine, including the theft of water which resulted from Petmin ignoring even relatively weak environmental regulations.

Tendele mine is the largest African producer of anthracite used in steel, an industry whose local output halved due to Chinese competition during the 2010s. Much more is now exported to Brazil. Tendele also provides several million tons of thermal coal annually for both the South African market – where nearly 90 percent of power is based on coal – and export markets, contributing to the climate catastrophe. In lieu of any capacity or political will to carry out a thorough-going full-cost accounting informed by environmental economics, and to apply polluter-pay principles, such global damage was never assessed in the coal firm's and government's environmental impact analyses.

Days after Ntshangase's assassination, Petmin's reply to critics reflected its confidence that the state's backing – and worker survival needs in these desperate circumstances – were sufficient to roll over objections. The Tendele coal mine, Petmin insisted,

> is the primary source of economic activity for a community of 220,000 people in the area 85km north of Richards Bay in KwaZulu-Natal. To be sustainable for more than another two years, the mine needs to expand its operations to be able to mine a new ore body which will extend the life of the operation by another 10 years. Mining rights for this have been duly granted. (Stoddard, 2020)

Indeed Petmin's three core points are true, since there are currently few employment alternatives; Tendele mine has voracious expansion plans; and the state suffers from mining industry influence, to the extent that Petmin had for years simply ignored water-use licensing requirement when *washing its coal for better combustion*, even though the Umfolozi River and its water table regularly suffered from droughts. For peasants and other residents, water shortages worsened to the point in June 2020 that during one community protest just south of Somkhele, a Tendele coal truck was stopped by three men and set alight, and although it was completely burned, the driver fortunately escaped (Zincome, 2020). (There was no record of who committed the sabotage.)

Another example of community dissent was a 2017 demonstration in which, the *Zululand Observer* reported, men and women:

> barricaded the entrance to the mine as a protest against a lack of jobs ... preventing vehicles, employees, contractors and visitors from entering. The group reportedly forced a truck to tip its coal and prevented others from leaving the mine with their loads. The company said these actions resulted in "material loss of production and substantial financial losses." Shortly after the group was released with a warning, the Global Environmental Trust issued a statement in which it said the mine had promised the men jobs. "These men were relocated by the mine in 2014 and promised jobs and training which never materialized. They are all unemployed. The men blocked the entrance to the mine and demanded to speak to mine management about their grievances. The police were called and the men were arrested. Later, the wives and mothers of these men did the same thing in protest of the earlier arrests and mine management had them arrested too."

Without substantial alternatives to mining or corporate agriculture – such as a more coherent food sovereignty option, plus public works projects to prepare Somkhele for more climate chaos – Sizani Ngubane and Fikile Ntshangase correctly feared that the various forms of environmental, economic, social and physical violence so common in South Africa would fuse against rural women. To illustrate, Tendele's own 'Local Economic Development' strategies in Somkhele were terribly weak even on their own terms, an in-house audit revealed (SRK Consulting, 2014). Accountability through state regulation and the courts was also impossible give the adverse balance of forces.

'The right to say no!'

Standing up against these realities and demanding a different future is therefore hazardous, due to the combined forces of state and capital. The assassination of Ntshangase was just one instance reflecting extreme dangers facing human rights and environmental defenders in KZN Province's many resource conflict sites (Corcoran, 2020; Heinrich Boell Foundation, 2019). As the Global Initiative Against Transnational Organized Crime (2019) had reported just five weeks before her murder,

> Since 2016, there have been at least 38 assassinations and 14 attempted assassinations in mining localities in KZN ... Our research found only one hit definitively linked to the KZN mining industry before 2016. Other organizations such as Human Rights Watch have documented the climate of fear and oppression which activists in South African mining communities endure.

Why the rise in tension in the period after 2015? It may not be a coincidence that the commodity super-cycle ended that year. Mineral prices had soared, starting in 2002, peaking just before the world financial meltdown in 2008, then recovering and plateauing from 2009 to 2014, and collapsing in 2015. For both the main mineral exports from KZN, coal and titanium dioxide, the peak dollar prices peaked in 2006–08, and have since zig-zagged, with coal's price going from $150/ton in 2008 to $44/ton in 2015, and titanium dioxide's from $2900 per metric ton in 2012 to a low of $1700 in early 2016. With price declines of that magnitude in 2015 (partially compensated by South Africa's own currency crash), the mining industry became both more oriented to increase volumes – so as to make up for lower prices per unit – and also more violent. The assassinations associated with titanium extraction by the Richards Bay Minerals (a subsidiary of Rio Tinto) were prolific, and in 2021 even the firm's local CEO was the victim of assassination.

The subsequent coal price explosion – partly due to the post-lockdown mini-boom and partly to the Russian invasion of Ukraine – is not encouraging, from the standpoint of preventing climate catastrophe. Across the world, thermal coal remains in exceptionally high demand: from $50/ton at the low point in mid-2020 to more than $400/ton two years later. German Chancellor Olaf Scholz visited Ramaphosa in Pretoria in May 2022 after Russia's Ukraine invasion, to request millions of tons in coal exports to Germany.

South African coal export price (Richards Bay), 2008–22

But resistance continues, so on the supply-side, anti-mining movements were important not only in Somkhele but in other coal sites. Causes of the conflict partly reflect shifts in global commodity prices and national politics, whether due to the mines' attempts at increasing output volumes to compensate during times of falling prices, or their failure to trickle-down benefits to communities during price increases. Another reflection of community protest occurred in mid-2021, when many coal mines (especially Tendele) were shuttered due to generalized political chaos in KZN province following the arrest of the former president Jacob Zuma (who hailed from Nkandla, a district close to Somkhele). Whichever theory holds, according to the Global Initiative Against Transnational Organized Crime (2019), in this part of South Africa,

> mines are frequently at the frontline of community frustration. The industry itself has noted increased opposition to its operations among neighboring communities. Incidents such as roadblocks, hijacking of vehicles to transport protesters, arson, damage to property and labor disruption have reportedly almost doubled in some areas between 2018 and 2019.

The philosophical stance of many mining-community activists is now often termed 'the Right to say No!' A vibrant alliance of movements is making this case not only at anti-mining protests, but on twitter – https://twitter.com/hashtag/right2sayno?f=live – and in the World Social Forum's (2020a, 2020b) activist webinars or its Thematic Forum on Mining and the Extractivist Economy. This movement gathered in 2018 at the first such major international Thematic Forum anti-mining event, which was in Johannesburg (Bond, 2018a). It contrasts vividly with the 2009 African Mining Vision – an African Union (2009) plan that celebrates transnational corporations – and the annual African Alternative Mining Indaba (Bond, 2015). One international solidarity network promoting this movement is Mines and Communities (2021).

Though the Mfolozi Community Environmental Justice Organization's members differed over whether to advocate closing Tendele outright, the *Zululand Observer* reported in 2016 that an important local group – the Mpukunyoni Community Property Association, led by local entrepreneur Bongani Pearce – regularly marched to that end. Their members were 'fed-up with living in close proximity to Somkhele mine' (Jolly, 2016). Pearce regularly suffered physical attacks and even arson due to his strong anti-mining stance. In 2018, neighbouring communities southwest of Somkhele, in Fuleni, engaged in successful protests against a proposed coal mine that would have bordered the Hluhluwe-iMfolozi Park (Jolly, 2018).

Following the money to the bankers, and then to Ohio

While activists and their lawyers are warning off the mining industry from pursuing eco-socially destructive projects like those at Somkhele and Xolobeni (both mining and offshore fossil fuel extraction), the firms' bankers are also receiving attention, stretching from national to global scale. In addition to Petmin's main local financier, Nedbank, one target is the World Bank, because it was one of Tendele's main funders when full-fledged mining began. The Bank's private sector investment arm, the International Finance Corporation (2002), was the original backer of the venture capital fund that resurrected Petmin – an apartheid-era shell company – at Somkhele.

The firm's subsequent primary banker, though, was South Africa's fourth largest lender, Nedbank, an institution which built its public relations strategy around being the country's most eco-aware financier. Yet in September 2007 it began lending to Petmin with a generous $11.5 million credit, most of which went to Somkhele for coal mining. In 2015 Nedbank offered a $19 million loan package to Petmin to make possible the 20 percent 'Broad-Based Empowerment' deal that ultimately split the community. Since Petmin's dividend payments covered only the interest payments, the passive 'owners' (local ethnic elites and labour leadership) were simply earning notional wealth, in turn leading to growing suspicions from those the coal mine oppressed: community members affected by pollution, ordinary workers and progressive conservationists.

As for Nedbank, it was a no risk proposition, because 'Petmin effectively guarantees the redemption of the Nedbank funding as it will subscribe for additional preference shares in the event of a potential default in payments' (Fin24, 2015). The deal is so pro-creditor that in mid-2020 when the Covid-19 lockdown briefly prevented Tendele from operating, and hence caused a loan repayment default, Nedbank gladly rolled it over. The interest rates paid in South Africa on commercial bank loans far exceeded the rate of inflation (usually by more than 6 percent); and internationally, very few governments typically paid a higher interest rate on 10-year state bonds. In this context, three weeks before Ntshangase's assassination, Nedbank (2020) issued a self-congratulatory press release: 'we are proud to say that in 2020 we have proven once again that green is the new gold standard'.

This combination of corporate public relations and corporate coziness between regulators and regulated financial institutions gave South Africa the leading position in PwC's (2020) biannual ratings of corporate corruption ('economic crime') during the 2010s. In 2019, a Treasury official admitted that 3–7 percent of the country's annual GDP output is squandered through Illicit Financial Flows (Planting, 2019). Whether licit or illicit, Petmin found a route to evacuate massive funding flows to the United States.

The Somkhele profits were, from 2019, used to begin construction on a $474 million Lake Erie (Ohio) pig-iron plant in Ashtabula, Ohio, near Cleveland. Similar to Somkhele's MCEJO, the Ashtabula community's leading environmental activists are upset with their state government's environmental regulator. The Ashtabula County Water Watch NGO appealed against state approval of Petmin's factory, especially once a 2020 filing of new – much higher – estimated emissions of nitrogen oxides, particulate matter and CO2 equivalents was made public. That meant, as the group complained,

> The Petmin project relies on natural gas produced in a way that is immensely taxing on the environment and public health. We may think the consequences of fracking are out of sight and mind in western Pennsylvania, but wastewater from this process is being pumped into the earth here in any of 15 active fracking wastewater injection wells in Ashtabula County. More wells will likely follow as the fracking industry continues to ramp up production since individual wells are becoming less productive and require more and more fracking fluid. Basically, we are cheaply selling clean water from Lake Erie to the fracking industry in Pennsylvania only to have it severely polluted and then shipped back to Ashtabula County and injected into the earth where it puts our water and soil resources at risk of contamination and causes increased risk of earthquakes. (Haytcher, 2020)

By mid-2021, Petmin's internal calculations appeared to have shifted, with the construction that had begun there in late 2020 suddenly halted a few weeks later, leading a local politician to confess he was not sure if the pig iron plant 'will ever come to fruition' (Haytcher, 2021). In May 2022, Doig admitted there were 'supply chain issues' and that $60 million had been invested but more financing was needed – but his hopes were buoyed by rising 'interest in the supply of U.S. manufactured pig iron' due to the Russian invasion of Ukraine (Terry, 2022).

The opportunity to derail Petmin would increase if Somkhele (or even Ashtabula) activism could be sustained. But it would also reflect the company's own financial contradictions. With Petmin's, 2020 corporate accounting disclosures, it was somewhat easier to assess how much Ntshangase and other local critics of coal mining were costing Tendele mine. At that stage, Tendele mine was running out of easy-to-dig surface coal and turned to drilling much deeper into the KZN crust, because – due especially to legal opposition and community protests – Petmin (2020) did not get easy approval for the new areas it wanted to stripmine.

Indeed the firm's legal dispute with MCEJO over an existing mining right and environmental authorization granted by the state in 2013 (and another giving three-decade mining permission from 2016 to 2046) stretched into 2022. On 4 May, a Pretoria High Court judge determined that the application and state approval were a 'breach of the law with regard to public participation', because consultation was limited to the Mpukunyoni Traditional Council. The mining house was, according to the judge, an 'unbridled horse that showed little to no regard for the law' (All Rise, 2022).

Petmin's (2022) reply, in contrast, welcomed the judge's 'pragmatic' finding that the mine did not need to close but instead that the Minister of Mineral Resources and Energy - who is notoriously pro-mining - must reconsider, leaving the *status quo* in force. And within two weeks, the mining house and ethnic leaders had gathered other allies - including two huge trade unions with more than 100,000 members each (AMCU and the National Union of Mineworkers) - to demand: 'All Rise and their funders to account to the community of their efforts to close Tendele Mine'. The strategy was to imply that it was outsiders to blame, a tactic also used in a different court (unsuccessfully) by a few breakaway MCEJO members earlier in the year (All Rise, 2022).

Petmin's profits, losses and capital depreciation remain opaque because the firm was, after 2017, no longer publicly-listed and hence not subject to shareholder oversight (its management removed it from the Johannesburg Stock Exchange in a buyout). Petmin's bookkeeper KPMG and the broader South African accountancy profession were seen as unreliable, if not outright corrupt, given KPMG's roles in various local corporate scandals even exceeding those of PwC, EY, Deloitte, McKinsey and Bain (Business Day, 2020; Dhlamini, 2017).

What Petmin (2020) did reveal, however, were massive 2018–20 machinery write-downs (capital depreciation) in Somkhele that did not correlate to the firm's fast-rising output. Also evident from Petmin's (2020) report is that since Tendele mine began operating in 2004, $800 million worth of coal was dug, reaching an annual peak of $120 million in 2018. Over the years 2013–19, Petmin boasted an average net profit rate of nearly 10 percent, in part because South African royalties and taxes are so low (only drawing down around 15 percent of Petmin's revenue). Petmin (2020) suffered a sharp drop in sales in the fiscal year through June 2020. Covid-19's lockdown meant two local anthracite customers - Glencore (the world's biggest commodity trader) and Samancor - as well as Brazil's Vale (the world's second largest mining house) all stopped buying for a few weeks in April-May. That reported $35 million loss in sales resulted in a $7 million net loss for the year 2019–20. It was in this context that MCEJO's blockage of further coal digging was seen by Petmin and its mineworkers as an existential threat.

This crisis, in turn, prevented mid-2020 loan repayments to Nedbank, which was typically making $2 million a year in interest revenues from its Petmin lending. Indeed, due to the Covid-19 crash and KPMG's generous assessment of the financial records, Petmin claimed to be unable to pay $41 million in debt, requiring rescheduling. Yet at the same time, Petmin USA prepared to begin construction on the $474 million pig iron plant in Ohio. The conclusion that could be drawn from the financial data was that reparations are due to the people of Somkhele and South Africa, for a full-cost accounting of resource depletion. But it is also useful to follow the money trail back even earlier.

World Bank reparations also overdue

The World Bank's International Finance Corporation (IFC) funded part of Tendele's start up during a critical phase when the Bank took equity ownership stakes in mining 'juniors', as small firms are called. This investment was made under the impression - stated by IFC officials Rashad Kaldany and Haydee Celaya (IFC, 2002) - that this would 'entrepreneurs who have previously had limited access to capital'. Such phraseology in post-apartheid South Africa suggests black empowerment. But as the lawyers working for Ntshangase and MCEJO - Tembeka Ngcukaitobi and

Kirsten Youens – told the Supreme Court of Appeal on 3 November 2020, the mining operation was illegal from the outset, lacking crucial government permissions to make such a destructive impact on the local environment and people (Masiteng, 2020). Since those authorizations were missing when mining operations began, a precedent case in the Constitutional Court (Allpay) allowed Ngcukaitobi to argue that effective remedial relief must follow a declaration of unlawfulness. He asked for a disclosure of profits made by Tendele coal mining since its origins, and that such profits acquired illegally – which include generous dividends sent to the IFC for several years – 'should be paid over to communities or other entities' (Masiteng, 2020).

At the point when Petmin began operating Tendele in 2004, its promises of community prosperity, black advancement and environmental responsibility – made when the IFC initially got involved – were effectively jettisoned. While remaining an all-white-male executive operation (with one black board member), Petmin's roots in apartheid mining evolved into fronting for Black Economic Empowerment operations. As a small mining shell company, Petmin had raised venture funds to open Tendele from the New African Mining Fund (NAMF), which was the recipient of a $5 million initial IFC commitment escalating to $30 million for its second stage (albeit aborted in 2014 during the global mining crash).

NAMF founder Doig soon also became Petmin's highest-profile leader. He began working as Chief Operating Officer there in 2006, and his personal page on the corporate website highlights the IFC investment that got Tendele up and running. Petmin initially received the majority of funds from NAMF for its main operation, Somkhele, and that stake, in turn, was the main reason the NAMF could declare an annual 39 percent profit rate at the time the IFC enjoyed its 6 percent stake in the fund. It is not known how much the IFC earned from the NAMF and the largest investment, Tendele coal mine, for data about its Petmin shares remain opaque, but it is likely to exceed $10 million. Another NAMF investment at the time was catastrophic: the Canadian firm Africo was victimized by the notorious Israeli tycoon Dan Gertler and his New York ally the Och-Ziff investment firm, as pointed out in a report by the NGO Rights and Accountability in Development (2019), which complained about the IFC's blasé attitude to mining industry corruption.

If such profits were returned, they could begin a process of financing the desperately-needed decarbonization of Somkhele. For the economically-impoverished local men supporting the coal mine, there remains no alternative in what for generations had been an apartheid Bantustan zone of economic despair, where before freedom in 1994, the only cash income came from migrant workers' remittances. By using potential reparations payments to link more closely the community, conservationists and potential climate-solidarity movements, a demand for a genuine Just Transition would potentially be capable of addressing the deep-seated conflict over coal mining. In this way, protecting the climate also would entail interconnected eco-socio-economic objectives in Somkhele: emissions-source mitigation, drought adaptation and other climate-crisis resilience projects, funded by the state, self-managed by workers, controlled by communities and especially women and youth. The need has never been greater.

Specifically, though, World Bank reparations are certainly due to oppressed South Africans. Apartheid profiteering by the Bank began in 1951 and lasted more than two decades before the United Nations forbade further loans. Its neoliberal policy advice is a major reason inequality increased to the world's highest level (a Gini Coefficient of 0.63) *after apartheid ended in 1994.* Other Bank and IFC investments here have reeked of socio-ecological destruction, the most predatory capitalism, and outright corruption. For example, the largest coal-fired power plant underway anywhere now was the Bank's largest-ever loan: $3.75 billion for Eskom's Medupi generator, one riddled with fraud and without exaggeration a clear case of 'Odious Debt', i.e. of such dubious origin that it should not be repaid (Bond, 2014).

At the same time the IFC was investing in the NAMF, another controversial mining stake in South Africa was simultaneously underway: the IFC's $50 million equity share in – and promised $100 million loan to – Lonmin, so as to promote Community Social Investment (CSI) at a particularly controversial mine producing a vast share of the firm's platinum: Marikana (Bond, 2018b). The IFC made this mine its poster child for CSI, yet the hatred that mineworkers and their community supporters felt towards Lonmin built up, resulting in the strike in August-September 2012, five years into the IFC intervention. In 2008–09, Ramaphosa – by then a major coal-mining tycoon – went into alliance with the world's largest commodity dealer, Glencore, and attempted to purchase Tendele coal mine. But they failed, and Petmin retained control (South African Competition Tribunal, 2009).

In 2015, the main Marikana women's group, Sikhala Sonke, attempted to get the World Bank's Compliance Advisor/Ombudsman (CAO) to force the IFC to take responsibility, requesting a formal Dispute Resolution process with Lonmin to give community relief from socio-economic repression. They gave up after the internal Bank process proved useless (Bond, 2018b). Lonmin, facing bankruptcy, was purchased by a local mining house in 2017 and the Sikhala Sonke grievances against the IFC and the Marikana mining operation remain unresolved. In another appeal to the Bank CAO to compel the IFC to make good on massive damages caused by a South African investment, the well-regarded women-led social advocacy group Black Sash (2017) insisted the IFC's $107 million (22 percent) share in Cash Paymaster Services compelled it 'to put measures in place to address and rectify impugned conduct', ultimately driving the firm into bankruptcy (Black Sash, 2020; Postman, 2020).

All of these efforts to discipline IFC-owned firms in South Africa occurred prior to the Bank's 2019 loss of its immunity from prosecution in the U.S. Supreme Court (Center for International Environmental Law, 2019). That may scare the Bank into settling in other jurisdictions, for fear the Foreign Corrupt Practices Act or other tort law may hold the IFC accountable in the U.S., where it has never had prior reason to fear prosecution. In the meantime, since the culture of criminality that prevails in the South African IFC portfolio is widespread, instead of relying upon the institution's own fatally-flawed internal review mechanisms, South Africa's courts may consider taking the advice of MCEJO's lawyers. That would entail beginning a long-overdue process of not only cease-and-desist against IFC-owned corporations, but the compulsion of reparations payments from the firms and their ethics-challenged financiers.

Conclusion: the climate reparations they demand and deserve

How to pay for all the Just Transition transformations that will be necessary in such sites? In addition to profit repatriation of the sort MCEJO's lawyers demanded from Petmin in 2019, another approach is a northern downpayment on climate debt due to places like the Somkhele villages, which were ravaged by a severe drought for several years. The concept of a wealthy northern country making such a downpayment comes from indigenous and eco-feminist Ecuadoreans who opposed Chevron (then Texaco) and later Chinese oil extraction from the eastern part of the country, in the upper reaches of the Amazon Basin. The Yasuni Park is the most biodiverse site in the world, but contained an estimated $10 billion of oil – so the 'Yasunization' strategy is the demand by climate justice activists that such fossil fuel *not be extracted*, but instead compensated (at $3.6 billion) for leaving unexploited (Temper et al., 2013). The proposal failed in

2013 because of bad faith negotiations by both the German and Ecuadoran governments. Berlin insisted on carbon trading instead of a climate debt downpayment to raise the needed funds, and Correa secretly negotiated Yasuni oil-drilling contracts with Sinopec as his Plan B. But the idea is continually revived.

The most famous proponents of Global North liability for climate damage are former Bolivian UN Ambassador Pablo Solon (Transnational Institute, 2010) and writer-activist Naomi Klein (2009). But a decade earlier, Quito-based eco-feminists Accion Ecologica (1999) and Joan Martinez-Alier were defining the concept as, 'the debt accumulated by the Northern industrial countries towards the countries and peoples of the South on account of resource plundering, environmental damages, and the free occupation of environmental space to deposit wastes, such as greenhouse gases'.

Conditions to establish a fund that could have helped Ntshangase and MCEJO activists fight Petmin more effectively were not yet in place (Bond, 2018c). On the one hand, the balance of forces were not likely to change, with former U.S. Secretary of State John Kerry – appointed the Biden Administration's climate czar in early 2021 – repeatedly unwilling to open up debate on the U.S. historic obligations for polluter-pays reparations. But on the other hand, Northern climate-liability consciousness would necessarily return, as many poor countries insisted that 'Loss & Damage' demands be returned to the 2022 United Nations climate summit in Egypt.

Activists, too, were remobilizing to remind of both ethical and economic-efficiency arguments for pricing carbon in a manner that would result in reparations, as witnessed in 350.org founder Bill McKibben's (2020) *New Yorker* column on 2 December 2020, entitled 'The climate debt the U.S. owes the world'. Paying reparations, McKibben concludes, is 'the only honorable, and only sensible, course: they don't call it global warming for nothing, and you can't control it anywhere without controlling it everywhere'. And if fighting fossil fuels is one of the most appropriate of beneficiaries of such reparations payments, especially under conditions in which Ntshangase's murder offers a permanent reminder of the gendered violence so characteristic of extractivism, then the Somkhele movement led by MCEJO is one of the most appropriate places in which to invest.

Disclosure statement

No potential conflict of interest was reported by the author(s).

ORCID

Patrick Bond http://orcid.org/0000-0001-6657-7898

References

Acción Ecológica. (1999). No more plunder, they owe us the ecological debt! *Bulletin of Acción Ecológica* 78.

All Rise. (2022, July). Open letter. https://allrise.org.za/open-letter-civil-society-organisations-call-on-attorney-dennis-sibuyi-and-his-clients-to-respect-the-constitutional-rights-of-community-members/

Bega, S. (2020, April 27). Village women's struggle for water worsened by Covid-19. *The Mercury*. https://www.iol.co.za/saturday-star/news/village-womens-struggle-for-water-worsened-by-covid-19-47191208

Berry, S. (2015, April 30). The fall of the wild. *Witness*. https://www.news24.com/witness/archive/The-FALL-of-the-wild-20150430

Black Sash. (2017, September 14). South African NGOs question IFC's investment into Net1. https://www.blacksash.org.za/index.php/media-and-publications/media-statements/81-south-african-ngos-question-ifc-s-investment-into-net1

Black Sash. (2020). Social Grants: Challenging Reckless Lending in South Africa. Cape Town. https://www.blacksash.org.za/images/publications/Social_Grants_-_Challenging_Reckless_Lending_in_South_Africa_FINALCHANGES_Thurs10092020.pdf

Bond, P. (2014). Marikana's meaning for crisis management. In U. Schuerkens (Ed.), *Global management, local resistances: Theoretical discussion and empirical case studies* (pp. 192–219). Routledge. https://www.taylorfrancis.com/books/e/9781315848341/chapters/10.4324978131584834192-219

Bond, P. (2015, March 14). Mining, energy, climate, capitalism: Why don't NGOs connect the dots? *Climate&Capitalism.* https://climateandcapitalism.com/2015/03/14/mining-energy-climate-capitalism-connect-the-dots/

Bond, P. (2018a, November 13). Mining conflicts multiply as critics of extractivism gather in Johannesburg. *CounterPunch.* https://www.counterpunch.org/2018/11/13/mining-conflicts-multiply-as-critics-of-extractivism-gather-in-johannesburg/

Bond, P. (2018b). From BASF to Volkswagen and the World Bank. In M. Grimm, J. Krameritsch, & B. Bekker (Eds.), *Business as usual after Marikana: Corporate power and human rights* (pp. 312–327). Jacana. https://rosalux.co.za/wp-content/uploads/2018/12/Business-As-Usual-2018.pdf

Bond, P. (2018c). Climate debt, community resistance and conservation alliances against KwaZulu-natal coal mining at Africa's oldest nature reserve. In B. Engels & K. Dietz (Eds.), *Climate change in Africa* (pp. 314–337). Peter Lang Verlag. https://www.peterlang.com/view/9783631742426/chapter-002.xhtml

Bond, P., Mdlongwa, T., & Zikhali, T. (2021). South African food politics. In N. Bohler-Muller, C. Soudien, & V. Reddy (Eds.), *Ethics, politics, inequality: New directions: State of the nation* (pp. 195–219). Human Sciences Research Council Press. https://www.hsrcpress.ac.za/books/ethics-politics-inequality-new-directions

Business Day. (2020, December 15). How a mighty profession has fallen. https://www.businesslive.co.za/bd/opinion/editorials/2020-12-15-editorial-how-a-mighty-profession-has-fallen/

Center for International Environmental Law. (2019, February 27). Supreme Court rules that World Bank Group can be sued in US courts in historic decision. https://www.ciel.org/news/supreme-court-rules-world-bank-group-immunity-jam-v-ifc/

Centre for Environmental Rights. (2020, October 24). Communities in Mpumalanga are demanding meaningful consultative forums to address the serious health impacts of water pollution caused by coal-mining companies. https://cer.org.za/news/communities-in-mpumalanga-are-demanding-meaningful-consultative-forums-to-address-the-serious-health-impacts-of-water-pollution-caused-by-coal-mining-companies

Cock, J. (2019). Resistance to coal inequalities and the possibilities of a just transition in South Africa. *Development Southern Africa*, *36*(6), 860–873. https://doi.org/10.1080/0376835X.2019.1660859

Corcoran, B. (2020, November 3). Murder of South African anti-mining activist highlights dangers for campaigners. *Irish Times.* https://www.irishtimes.com/news/world/africa/murder-of-south-african-anti-mining-activist-highlights-dangers-for-campaigners-1.4399212

Dhlamini, S. (2017, October 4). Wits ditches KPMG. *Polity.* https://www.polity.org.za/article/wits-ditches-kpmg-2017-10-04

Doig, B. (2016, March 8). Interview. *Business Day.* https://www.businesslive.co.za/bd/companies/mining/2016-03-08-business-day-tv-diversified-miner-petmin-weathers-the-drought/

Fin24. (2015, June 10). Petmin announces R350m empowerment deal. *News24.* https://www.news24.com/Fin24/petmin-announces-r350m-empowerment-deal-20150610

Global Initiative Against Transnational Organized Crime. (2019, September 14). eMpembeni mirrors the scourge of mining-related assassinations in KZN. https://www.dailymaverick.co.za/article/2020-09-14-empembeni-mirrors-the-scourge-of-mining-related-assassinations-in-kzn/

Haytcher, B. (2020, August 28). Group appeals Petmin permit. *Star Beacon.* https://www.starbeacon.com/news/local_news/group-appeals-petmin-permit/article_c5450fc1-fa26-5a50-936f-e4bf3bd8269f.html

Haytcher, B. (2021, June 18). Ashtabula City Council Ward 4 candidate discusses issues. *Star Beacon.* https://www.starbeacon.com/news/local_news/ashtabula-city-council-ward-4-candidate-discusses-issues/article_730c262e-b3d9-5aba-af9d-9dcd157ec78e.html

Heinrich Boell Foundation. (2019, May 31). Uncalculated risks. https://www.boell.de/en/2019/05/31/uncalculated-risks-threats-and-attacks-against-human-rights-defenders-and-role

International Finance Corporation. (2002, August 29). IFC provides financing to junior mining sector in South Africa. https://ifcext.ifc.org/IFCExt/Pressroom/IFCPressRoom.nsf/0/91AF43A22691289285256C240055F0B2?OpenDocument

Jolly, T. (2016, June 1). Somkhele communities seek mine closure: Fed-up with living in close proximity to Somkhele mine, Mpukunyoni communities seek closure of the mine. *Zululand Observer*. https://zululandobserver.co.za/114064/somkhele-communities-seek-mine-closure/

Jolly, T. (2018, June 21). New mining threat looms. *Zululand Observer*. https://zululandobserver.co.za/173920/new-mining-threat-looms/

Klein, N. (2009, November 23). Why rich countries should pay reparations to poor countries for the climate crisis. *Democracy Now*. https://www.democracynow.org/2009/11/23/naomi_klein_on_climate_debt_why

Kockett, F., & Hattingh, M. (2020, October 28). South African activist killed as contentious coal mine seeks to expand. *Mongabay*. https://news.mongabay.com/2020/10/south-african-activist-killed-as-contentious-coal-mine-seeks-to-expand/

Madhanagopal, D., Bond, P., & Bayón, M. (2022). Eco-feminisms in theory and practice in the global south: India, South Africa, and Ecuador. In D. Madhanagopal, C. Beer, B. Nikku, & A. Pelser (Eds.), *Environment, climate, and social justice: Perspectives and practices from the global south* (pp. 275–296). Springer Nature.

Masiteng, M. (2020, November 3). Tendele Coal Mining should be ordered to declare its profits: Ngcukaitobi. South African Broadcasting Corporation. https://www.sabcnews.com/sabcnews/tendele-coal-mining-should-be-ordered-to-declare-its-profits-ngcukaitobi/

McKibben, B. (2020, December 2). The climate debt the U.S. owes the world. *New Yorker*. https://www.newyorker.com/news/annals-of-a-warming-planet/the-climate-debt-the-us-owes-the-world

Mfolozi Community Environmental Justice Organization (and 15 other groups). (2020, October 23). The killing of Somkhele environmental activist Fikile Ntshangase: A joint statement. https://cer.org.za/news/the-killing-of-somkhele-environmental-activist-fikile-ntshangase-a-joint-statement

Mines and Communities. (2021). http://www.minesandcommunities.org/

Modjadji, N., & Goba, N. (2018, May 15). 'Apartheid was not a crime against humanity': AfriForum CEO's remark sparks outrage. *TimesLive*. https://www.timeslive.co.za/politics/2018-05-15-uproar-over-remark-that-apartheid-was-not-a-crime-against-humanity-by-afriforum-ceo/

Msimang, S. (2018, August 28). South Africa's Afriforum and Trump's kiss of death. *Al Jazeera*. https://www.aljazeera.com/opinions/2018/8/28/south-africas-afriforum-and-trumps-kiss-of-death

National Union of Metalworkers of South Africa. (2022). 11th National Congress Secretariat Report 15–19 June, Johannesburg.

Nedbank. (2020, October 1). Experts who drive impactful outcomes. https://www.nedbank.co.za/content/nedbank/desktop/gt/en/news/corporate-and-investment-banking-news/press-releases/2020/winner-- env ironmental-finance-impact-awards.html

Ngema, T. (2020, November 26). Seven white rhino carcasses found in eight days at Hluhluwe-iMfolozi Park. *Daily News*. https://www.iol.co.za/dailynews/news/seven-white-rhino-carcasses-found-in-eight-days-at-hluhluwe-imfolozi-park-dfd40892-b4bb-47c1-82eb-ab26240e6d25

Petmin. (2015, June 10). Petmin announces R350 million Broad-Based Black Economic Empowerment (BBBEE) deal. http://www.sharedata.co.za/sensad.asp?id=252157

Petmin. (2020, November 17). Somkhele Anthracite Mine: Update on Social and Labor Plans (SLP) – 19 November 2020. http://petmin.co.za/pdf/somkhele-social-and-labor-plans/slp-update-19-11-2020.pdf

Petmin. (2022, May 8). Tendele mining right maintained as judge calls for pragmatism Johannesburg. https://www.petmin.co.za/pdf/media-release/Tendele-Mining-right-maintained-as-judge-calls-for-progmatism.pdf

Pitt, B. (2020, November 19). The struggle for Mfolozi and the murder of Fikile Ntshangase: Who is blackmailing who? *Daily Maverick*. https://www.dailymaverick.co.za/article/2020-11-19-the-struggle-for-mfolozi-and-the-murder-of-fikile-ntshangase-who-is-blackmailing-who/

Planting, S. (2019, October 27). SA's anti-money laundering measures under global spotlight. *Daily Maverick*. https://www.dailymaverick.co.za/article/2019-10-27-sas-anti-money-laundering-measures-under-global-spotlight/

Postman, Z. (2020, September 23). "Insolvent" Cash Paymaster Services must still pay back profits to SASSA, court told. *GroundUp*. https://www.groundup.org.za/article/insolvent-cash-paymaster-services-must-still-pay-back-profits-to-sassa-court-told/

PwC. (2020, February 1). Global economic crime and fraud survey 2020. https://www.pwc.co.za/en/publications/global-economic-crime-survey.html

Rights and Accountability in Development. (2019, September 17). IFC's ill-judged DR Congo Mines: Why has IFC abandoned Congolese victims of corruption? http://www.raid-uk.org/sites/default/files/ifcs_ill-judged_dr_congo_mining_projects_-_full_report_final_.pdf

Save Our Wilderness. (2021). Films. https://saveourwilderness.org/films/

South African Competition Tribunal. (2009, May 6). Shanduka Coal (Pty) Ltd v Springlake Holdings (Pty) Ltd. http://www.saflii.org/za/cases/ZACT/2009/27.html

South African Food Sovereignty Campaign. (2021). Water stressed communities map. Johannesburg. https://www.safsc.org.za/water-stressed-communities-map/

SRK Consulting. (2014, March 4). A competent person's report on the material assets of Tendele Coal Mining (Pty) Ltd. http://www.overend.co.za/download/Tendele_Coal_Mining.pdf

Stoddard, E. (2020, November 19). Petmin defends its resettlement plan for controversial Somkhele mine. *Business Maverick*. https://www.dailymaverick.co.za/article/2020-11-19-petmin-defends-its-resettlement-plan-for-controversial-somkhele-mine/

Temper, L., Yánez, I., Sharife, K., Ojo, G., & Martinez-Alier, J. (2013, May 21). Towards a post-oil civilization. Yasunization and other initiatives to leave fossil fuels in the soil. *Ejolt report #6*. http://www.ejolt.org/2013/05/towards-a-post-oil-civilization-yasunization-and-other-initiatives-to-leave-fossil-fuels-in-the-soil/

Terry, S. (2022, May 16). Petmin project still on. *Star Beacon*. https://www.starbeacon.com/news/local_news/petmin-project-still-on/article_25fb1a9c-d2c1-11ec-a53c-833070a910b5.html

Transnational Institute. (2010). Bolivia provides resistance and hope at Brokenhagen. https://www.tni.org/es/node/13008

Walters, T. (2021, January 3). Our Burning Planet hero of the year: Fikile Ntshangase. *Daily Maverick*. https://www.dailymaverick.co.za/article/2021-01-03-our-burning-planet-hero-of-the-year-fikile-ntshangase/

Wildlife and Environment Society of South Africa. (2014, June 5). WESSA objection against Fuleni anthracite coal mine application. https://www.engineeringnews.co.za/article/essa-objection-against-fuleni-anthracite-coal-mine-application-2014-06-05

WoMin. (2017, February 24). No longer a life worth living. https://womin.africa/download/wominparticipatoryactionreport_english/?wpdmdl=4996&refresh=5fc64c385b5a31606831160

World Social Forum. (2020a). Global Webinar Series. https://www.thematicsocialforum.org/globalwebinarseries

World Social Forum. (2020a). Thematic Social Forum. https://www.thematicsocialforum.org/

Zincome, M. (2020, June 8). Coal truck petrol bombed in Mtubatuba: The incident occurred during an alleged protest over water shortages. *Zululand Observer*. https://zululandobserver.co.za/226573/coal-truck-petrol-bombed-in-mtubatuba/

Index

Page numbers in **bold** refer to tables and those in *italic* refer to figures.